Introduction to R for Social Scientists

Chapman & Hall/CRC
Statistics in the Social and Behavioral Sciences Series

Series Editors
Jeff Gill, Steven Heeringa, Wim J. van der Linden, Tom Snijders
Recently Published Titles

Multilevel Modelling Using Mplus
Holmes Finch and Jocelyn Bolin

Bayesian Psychometric Modeling
Roy Levy and Robert J. Mislevy

Applied Survey Data Analysis, Second Edition
Steven G. Heering, Brady T. West, and Patricia A. Berglund

Adaptive Survey Design
Barry Schouten, Andy Peytchev, and James Wagner

Handbook of Item Response Theory, Volume One: Models
Wim J. van der Linden

Handbook of Item Response Theory, Volume Two: Statistical Tools
Wim J. van der Linden

Handbook of Item Response Theory, Volume Three: Applications
Wim J. van der Linden

Bayesian Demographic Estimation and Forecasting
John Bryant and Junni L. Zhang

Multivariate Analysis in the Behavioral Sciences, Second Edition
Kimmo Vehķalahti and Brian S. Everitt

Analysis of Integrated Data
Li-Chun Zhang and Raymond L. Chambers

Multilevel Modeling Using R, Second Edition
W. Holmes Finch, Joselyn E. Bolin, and Ken Kelley

Modelling Spatial and Spatial-Temporal Data: A Bayesian Approach
Robert Haining and Guangquan Li

Handbook of Automated Scoring: Theory into Practice
Duanli Yan, André A. Rupp, and Peter W. Foltz

Interviewer Effects from a Total Survey Error Perspective
Kristen Olson, Jolene D. Smyth, Jennifer Dykema, Allyson Holbrook, Frauke Kreuter, and Brady T. West

Measurement Models for Psychological Attributes
Klaas Sijtsma and Andries van der Ark

Big Data and Social Science: Data Science Methods and Tools for Research and Practice, Second Edition
Ian Foster, Rayid Ghani, Ron S. Jarmin, Frauke Kreuter and Julia Lane

Understanding Elections through Statistics: Polling, Prediction, and Testing
Ole J. Forsberg

Analyzing Spatial Models of Choice and Judgment, Second Edition
David A. Armstrong II, Ryan Bakker, Royce Carroll, Christopher Hare, Keith T. Poole and Howard Rosenthal

Introduction to R for Social Scientists: A Tidy Programming Approach
Ryan Kennedy and Philip Waggoner

For more information about this series, please visit: https://www.routledge.com/Chapman--Hall-CRC-Statistics-in-the-Social-and-Behavioral-Sciences/book-series/CHSTSOBESCI

Introduction to R for Social Scientists
A Tidy Programming Approach

by
Ryan Kennedy
Philip Waggoner

CRC Press
Taylor & Francis Group
Boca Raton London New York

CRC Press is an imprint of the
Taylor & Francis Group, an **informa** business

A CHAPMAN & HALL BOOK

First edition published 2021
by CRC Press
6000 Broken Sound Parkway NW, Suite 300, Boca Raton, FL 33487-2742

and by CRC Press
2 Park Square, Milton Park, Abingdon, Oxon, OX14 4RN

Library of Congress Cataloging-in-Publication Data

ISBN: 9780367460709 (hbk)
ISBN: 9780367460723 (pbk)
ISBN: 9781003030669 (ebk)

Typeset in Computer Modern font
by KnowledgeWorks Global Ltd.

Contents

Preface

This book is a distillation of our approach to programming in R for exploring and explaining a variety of social science behavior. It is the product of our own notes from teaching R to many groups of people, from undergraduates and graduates to faculty members and practitioners. Indeed, this book would be impossible without the support from and engagement by many students, colleagues, and folks generally interested in our work. We are deeply grateful to these people and are excited to share our work in this form.

In this book, we have two primary goals:

1. To introduce social scientists, both in and out of academia, to R. R is at the same time a programming language as well as an environment to do statistics and data science. As R is open source (meaning open contribution of packages via the Comprehensive R Archive Network (CRAN)), there are many powerful tools available to users in virtually any discipline or domain to accomplish virtually any statistical or data science task. Our goal, then, is to cover the tools we find most helpful in our research as social scientists.

2. As the subtitle of the book suggests, we are interested in exposing social scientists to the "tidy" approach to coding, which is also referred to as the *Tidyverse*. Though we expound on this in much greater detail throughout, the Tidyverse is a collection of packages all built around consistency and making tasks in R streamlined, with the product being a clean, clear rendering of the quantity or object of interest. And as this is an introductory text, we suggest it is most valuable to start from the Tidyverse framework, rather than base R, to reduce the steepness of the learning curve as much as possible.

Overview of Chapters

In the book, we cover the following topics for a full introduction to tidy R programming for social scientists:

1. **Introduction**: Motivation for the book, getting and using R
2. **Foundations**: Packages, libraries, and object-oriented programming
3. **Data Management**: Getting your data into workable, tidy form
4. **Visualization**: Visual presentations using `ggplot2` and the grammar of graphics
5. **Essential Programming**: Interacting with base R to learn functional programming
6. **Exploratory Data Analysis**: Exploring relationships and data in the Tidyverse
7. **Essential Statistical Modeling**: Fitting and diagnosing widely used models in the Tidyverse
8. **Parting Thoughts**: Conclusion and wrap-up

Acknowledgements

Though the final product we present in this book is our own (wherein we accept full responsibility for any errors), we could not have produced this book without the help and influence of many other excellent social scientists and programmers. Thus, in the same open-source spirit, we would like to acknowledge the following people for sharing and/or making code available: Ling Zhu, Scott Basinger, Thomas Leeper, Max Kuhn, and Hadley Wickham.

About the Authors

Dr. Ryan Kennedy is an associate professor of political science at the University of Houston and a research associate for the Hobby Center for Public Policy. His work in computational social science has dealt with issues of forecasting elections, political participation, international conflicts, protests, state failure, and disease spread. He has published in *Science*, the *American Political Science Review*, *Journal of Politics*, and *International Studies Quarterly*, among others. These articles have won several awards, including best paper in the *American Political Science Review*, and have been cited over 1,700 times since 2013. They have also drawn attention from media outlets like Time, The New York Times, and Smithsonian Magazine, among others. For more information, visit https://ryanpkennedy.weebly.com/.

Dr. Philip Waggoner is an assistant instructional professor of computational social science at the University of Chicago and a visiting research scholar at the Institute for Social and Economic Research and Policy at Columbia University. He is also an Editorial Board Member at the *Journal of Mathematical Sociology* and an Associate Editor at the *Journal of Open Research Software*, as well as a member of *easystats* (a software group that tends to an ecosystem of packages making statistics in R easy). He is the author of the recent book, *Unsupervised Machine Learning for Clustering in Political and Social Research* (Cambridge University Press, 2020). And in addition to authoring and co-authoring several R packages, his work has appeared or is forthcoming in numerous peer-reviewed journals including the *Journal of Politics*, *Journal of Mathematical Sociology*, *Journal of Statistical Theory and Practice*, *Journal of Open Source Software*, among others. For more information, visit https://pdwaggoner.github.io/.

1

Introduction

R is a widely used statistical environment that has become very popular in the social sciences because of its power and extensibility. However, the way that R is taught to many social scientists is, we think, less than ideal. Many social scientists come to R after learning another statistical program (e.g. SAS, SPSS, or Stata). There are a variety of reasons they do this, such as finding there are some tasks they cannot do in these other programs, collaborating with colleagues who work in R, and/or being told that they need to learn R. For others, R may be the first statistical program they encounter, but they come to it without any kind of experience with programming (or even, increasingly, using a text interface).

This is part of why "learning R" can be frustrating. Learning R for the first time, most students are shown how to undertake particular tasks in the style of a cookbook (i.e., here is how you conduct a regression analysis in R), with little effort dedicated to developing an underlying intuition of how R works as a language. As a result, for those who have experience with other statistical programs, R comes across as a harder way to do the same things they can do more easily in another program. This cookbook approach can also produce frustration for those who are coming to R as their first statistical analysis environment. Working with R in such a way becomes a process of copying and pasting, with only a shallow understanding of why things have a particular structure and, thus, difficulty moving beyond the demonstrated examples.

Finally, the cookbook approach is, in many ways, a holdover from the pre-internet era, when large coding manuals were a critical reference for finding out how to do anything in a complex program. These books had to be exhaustive, since they were needed as much for reference as for learning the environment. Today, however, there is a plethora of online materials to demonstrate how to perform specific tasks in R, and exhaustiveness can come at a cost to comprehension. What most beginners with R need is a concrete introduction to the fundamentals, which will allow them to fully leverage the tools available online.

This book is focused on equipping readers with the tools and knowledge to overcome their initial frustration and fully engage with R. We introduce a modern approach to programming in R – the Tidyverse. This set of tools introduces a consistent grammar for working with R that allows users to

1

quickly develop intuitions of how their code works and how to conduct new tasks. We have found this increases the speed of learning and encourages creativity in programming.

This book is based on an intensive 3-day workshop introducing R, taught by one of the authors at the Inter-University Consortium for Political and Social Research (ICPSR), as well as numerous workshops and classes (at both the undergraduate and graduate levels) conducted by both authors. The goal is to have the reader: (1) understand and feel comfortable using R for data analysis tasks, (2) have the skills necessary to approach just about any task or program in R with confidence, and (3) have an appreciation for that which R allows a researcher to do and a desire to further their knowledge.

1.1 Why R?

If you have picked up this book, chances are that you already have a reason for learning R. But let's go through some of the more common reasons why conducting your research in R is a good idea.

One of the major attractions of R is that it is free and open source. R was created by Ross Ihaka and Robert Gentleman, of the Department of Statistics at the University of Auckland, in the early 1990s (Ihaka and Gentleman, 1996). It was designed to be a dialect of the popular S-PLUS statistical language that was developed for Bell Labs. Unlike S-PLUS, however, R was released under the GNU General Public License, which allows users to freely download, alter, and redistribute it.

The result of this open source license is that R is accessible to everyone, without exorbitant licensing fees. It is also regularly updated and maintained, with frequent releases that allow for quick fixing of bugs and the addition of new features.[1] Perhaps most importantly, the open source nature allows users to contribute their own additions to R in the form of "packages." You will often hear R users say, in response to a question about how to do something in R, "There is a package for that." From running advanced statistical models to ordering an Uber (the ubeR package) or making a scatterplot with cats instead of points (the CatterPlots package), it is likely that someone has developed a way to do it in R. As of 2015, there were over 10,000 packages on the Comprehensive R Archive Network (CRAN), with scores more being created all the time. Indeed, the book you are reading now was originally written completely in R using R Markdown and the bookdown package (Xie, 2019).

---

[1]The major new release usually comes around October, so you should, at a minimum, update your R system around this time.

Another reason for learning R is flexibility. R is both a language and an "environment" where users can do statistics and analysis. This covers a lot of ground – from data visualization and exploratory data analysis, to complex modeling, advanced programming and computation. R allows you to scrape data from websites, interact with APIs, and even create your own online ("Shiny") applications. This flexibility, in turn, allows you as a researcher to undertake a wider variety of research tasks, some of which you might not even have considered previously.

Though R is wonderfully flexible, fast, and efficient, the learning curve can be quite steep, as users must learn to write code. For example, in some other popular statistics programs, users can point-and-click on the models they want with little to no interface with the mechanics behind what is going on. This is both good and bad. It is good in that the learning curve in point-and-click interfaces is much gentler and accommodating. However, it is not a great thing in that it restricts user interface with the process of coding and statistical analysis. Point-and-click encourages minimal interaction with the data and tasks, and ultimately following the well-trod path of others, *rather* than creating your own path.

The coding process required by R is also increasingly becoming the standard in the social sciences. The "replication revolution" in the social sciences has encouraged/required scholars to not only think about how they will share their results, but also how they will share the way they got those results (King, 1995; Collaboration et al., 2015; Freese and Peterson, 2017). Indeed, several of the top social science journals – including *American Economic Review, Journal of Political Economy, PLOS ONE, American Journal of Political Science*, and *Sociological Methods and Research*, among others – now require submission of replication code and/or data prior to publication. Still others strongly encourage the submission of replication code. R code is ideal for this purpose – there are almost no obstacles to other scholars downloading and running your R code. The same cannot be said about programs that require licenses and point-and-click interaction.

This replication process can also be useful for your own work. There is a common refrain among computer programmers that, "If you do not look at your code for a month, and have not included enough comments to explain what the commands do, it might as well have been written by someone else." The same is true of point-and-click software. If you have a process that is reasonably complex and you do not work with it for a while, you might completely forget how to do it. By writing an R script, you have a written record of how you did each task, which you can easily execute again.

Additionally, we recommend the use of R in a variety of applied research settings because of the high-quality options for visualization. Broadly, R uses layers to build plots. This layering provides many flexible options for users to interact directly with their visual tools to produce high-quality graphical depictions of quantities of interest. Further, some packages, e.g., `ggplot2`, use something called the "grammar of graphics" (Wilkinson, 2012), which is a process of streamlining the building of sophisticated plots and figures in R (Wickham et al., 2019b; Wickham, 2009; Healy, 2018). This and other similar packages offer users even more advanced tools for generating high-quality, publication-ready visualizations (Lüdecke et al., 2020).

And finally, we highly recommend R, because of the community. From blogs and local "R User" community groups in cities throughout the world to a host of conferences (e.g., UseR, EARL, rstudio::conf), the R community is a welcoming place. Further, the open source nature of R contributes to a communal atmosphere, where innovation and sophistication in programming and practice are highly prioritized. Put simply, R users want R to be the best it can be. The result is an inclusive community filled with creative programmers and applied users all contributing to this broader goal of a superior computing platform and language. And in the words of one of the most influential modern R developers, Hadley Wickham (a name you will see a lot in this book) (Waggoner, 2018a),

> . . . When you talk about choosing programming languages, I always say you shouldn't pick them based on technical merits, but rather pick them based on the community. And I think the R community is like really, really strong, vibrant, free, welcoming, and embraces a wide range of domains. So, if there are people like you using R, then your life is going to be much easier.

Therefore, though tricky to learn, if users are engaged in any way with data, whether working for an NGO, attending graduate school, or even legal work in many cases, users will be glad they opted to begin in R and endured the hard, but vastly rewarding work up front.

1.2 Why This Book?

There are many good introductions to R (Monogan III, 2015; Li, 2018; Wickham and Grolemund, 2017), and we will point you towards several of them throughout. Yet, this book provides a unique and beneficial starting place, particularly for social scientists. There are several features of this book that lead us to this conclusion.

First, it is written specifically for social scientists. Many of the best introductions to R are written for those who are coming from other programming languages (e.g. Python, C++, Java) or from database design (e.g., Spark, SQL). The assumption is that the reader will already be pretty familiar with programming concepts, like objects, functions, scope, or even with R itself. This, however, does not apply to most social scientists, who usually do not come in with experience in either programming or database management, and will, therefore, find these concepts unfamiliar, and often quite vexing. We also include details that are likely to be particularly relevant to social scientists, such as how to automatically generate tables using R.

Second, we write this as a genuine introduction course, *not* as a cookbook. Cookbooks have their place for learning R. They provide handy guides to completing particular tasks, and are indispensable as you go through your work. But, just as following the steps to make bread is not the same as understanding how bread is made, copying code from a book or online resource is not the same as developing the skill base to flourish as a data analyst who uses R. For a similar reason, unlike some other introductions, we do not create any special software specifically for this book – you are here to learn R, not a software we design. This book concentrates on helping you to understand what you are doing and why. After working your way through this book, you should be able to undertake a range of tasks in R and more easily learn new ones and even troubleshoot your own errors.

Third, we provide a thoroughly *modern* introduction to R. While using the word "modern" in any book is a risky proposition, we mean this in terms of using the latest tools as of this writing to help you be as productive as possible. This means using the RStudio integrated development environment (IDE) to assist you in writing and running code, R projects to keep track of and organize your work, and the Tidyverse set of tools to make your code more modular and comprehensible.

Fourth, we concentrate on the areas of learning R that you will use the most often and are typically the most frustrating for beginners. Many people have heard of the "Dunning-Kruger effect", which is the tendency for people with low ability to overestimate their ability (Kruger and Dunning, 1999). Many people forget about the inverse part of the Kruger-Dunning effect – the tendency for experts to underestimate the difficulty of tasks for which they are an expert. This sometimes exhibits itself in R introductions that attempt to introduce quite advanced statistical models, but give little to no attention to issues like file systems and data management. Yet, things like setting working directories are some of the most common stumbling blocks for students and data scientists will often say that 80% of their job is managing and shaping data, but this is almost never reflected in introductory texts. We try to correct this by giving a significant focus in the beginning to these fundamental skills.

Fifth, this is a very concise introduction to R. We do not intend to cover, nor do you need to know, everything about the internal workings of R or all of the different options and functions in the Tidyverse. For a working social scientist, the goal is to learn the parts you are likely to use most often, and gain enough understanding of how R functions to get help with the unique situations. We would argue that many "introductions" suffer from too much detail, where what is important is buried under an avalanche of options you are unlikely to use and will promptly forget.

Finally, the analysis demonstrated throughout is based on real survey data from the American National Election Study (ANES) and a large cross-country data set created by Professor Pippa Norris at Harvard University. We selected these examples because they are real social science data, thus allowing users to get their hands dirty in a very practical way, mirroring many contexts they are more likely to see in their future work, rather than conducting a demonstration using either pre-cleaned data or data about which the reader will have no intuition. We have not cleaned or processed the data in any way, so it provides a good example of what you will encounter in the "real world." You will be using the exact same data you would get from downloading these data sets from the internet.

1.3   Why the Tidyverse?

As was mentioned in the last section, this book is somewhat unique among social science introductions in our reliance on tools related to the "Tidyverse." This is a set of tools that have been collected and curated to make your work in R more productive. The Tidyverse is actually a collection of R packages (which we will discuss later), which all share an underlying design philosophy, grammar and structure (Wickham, 2017; Wickham and Grolemund, 2017; Wickham et al., 2019a).

There are several reasons we prefer to concentrate on the Tidyverse. First, it will allow us to get started with real data analysis, quickly. For those who do not start with a programming background, one of the more intimidating things about R is the introduction of programming concepts that usually comes with it. The basis of the R language was designed for Bell Labs engineers more than 50 years ago. The Tidyverse grammar was designed for data analysts from a wide range of backgrounds. The tools in the Tidyverse allow you to start getting meaningful data analysis right away.

Just as importantly, the shared design strategy of Tidyverse packages means that you will have an easier time learning how to do new things. The consistent design means that the intuitions you develop in this book should serve you well as you use new functions in the Tidyverse, allowing you to expand your knowledge more quickly.

Second, the Tidyverse grammar is more comprehensible for people coming from other statistical packages. The use of characters like $ or [[ ]] is often one of the most intimidating parts of learning R for beginners. We will learn these things in this book, but we will only do so after learning a range of consistent and simple functions that will achieve the main tasks you wish to accomplish in data analysis.

Third, the Tidyverse usually has a single obvious method for achieving a goal. This draws from the philosophy that there should be one, and preferably only one, obvious way to do a task. This is very useful for being able to learn quickly and to understand what is being done in any example. A simple illustration of this is creating a new variable that is our original variable times 1,000. In base R, there are at least three ways to do this.

```
dataset$new_variable <- dataset$old_variable * 1000

dataset[["new_variable"]] <- dataset[["old_variable"]] * 1000

dataset[, "new_variable"] <- dataset[, "old_variable"] * 1000
```

Since there is no right way to do it, you will often find different preferences within the same group of scholars (and sometimes within the same code). In contrast, there is only one way to create this variable in the Tidyverse:

```
dataset <- dataset %>%
  mutate(new_variable = old_variable * 1000)
```

From our experience, this makes it much easier to keep track of what is being done, share your code with others, and avoid frustration spending hours finding out what may have gone wrong with your analysis.

Fourth, while this book is intended primarily as an introduction to R, those who already know some R will find it useful for learning how to write "tidy" code in R. Many utilities in R are moving towards the Tidyverse structure and grammar, and this book will provide the familiarity with the Tidyverse needed to leverage these tools effectively.

Finally, the tools provided in the Tidyverse are extremely powerful. The `ggplot2` package, for example, has become the standard for most data visualization in R. The use of a consistent grammar makes it much easier to extend and develop than traditional R packages. Think about it like learning a foreign language. If the rules about, for example, how nouns are used changes from situation to situation, this makes it more difficult to learn the language and create your own statements. If, on the other hand, there is consistency in the rules, you can apply those rules to extend to new situations much more easily.

1.4 What Tools Are Needed?

Hopefully, if you are reading this book and have made it to this point, you are sufficiently convinced of R's value for both programming and statistical analysis, if you were not already convinced. So at this point we transition slightly away from a high-level discussion of R, and toward more practical aspects of how to get started with R.

1.4.1   Downloading R and RStudio

Before getting to into the environment, we first need to introduce precisely how to access the environment. It is undeniably daunting to open up R for the first time, and see a blank screen. Thus, we highly recommend the use of RStudio, which is a user-friendly integrated development environment (IDE) that directly interacts with the language of R (RStudio Team, 2015). In RStudio you can do all kind of things, from practicing writing code before running it (e.g., scripts), to writing reports (e.g., using markdown), to hopefully someday developing and releasing your own R packages. All of these and more are possible within RStudio directly.

Moreover, RStudio includes a number of useful features that make working in R easier. This includes automatically closing parentheses and brackets for you, highlighting which closed parentheses correspond with a particular open parenthesis, providing hints on how to complete the command you are writing, color-coding of your scripts, providing keyboard shortcuts for common commands, and marking likely errors. Other than perhaps the Emacs Speaks Statistics (ESS) suite for Emacs (which we do not recommend unless you already use and like Emacs), RStudio is the most complete IDE for users of R.

So where and how can users get R and RStudio? As mentioned earlier, perhaps one of the best things about R is that it is free. Users simply need to go to the R-Project page, http://www.r-project.org/, to first download R. Then, once R is successfully installed, go to the RStudio page, http://rstudio.com, to download RStudio onto your machine. For step-by-step download and installation procedures, with illustrations, you can go to the companion website, https://i2rss.weebly.com.

For those of you who have already downloaded and installed R and RStudio, we recommend you take some time to check whether you are working with the latest version, and, if not, to update both systems. To check your version of R, simply run the command `,version` (or for those of you working at the command line, `R --version`).[2] To check your version of RStudio, simply run the command, `rstudioapi::getVersion()` to return only the version number, or `rstudioapi::versionInfo()` to get the version number, mode (desktop, cloud, etc.), and the citation format for properly citing the use of RStudio.[3]

---

[2]A simpler tip is, when you open a new RStudio session, the version of R currently running will appear automatically in the open console window.

[3]Note that when reading a version number for any software, the first value is the major release, the middle value is more minor release, and subsequent values (3rd and 4th, e.g., "9000") reflect the most minor changes to the software. For example, upon writing this book, the latest release of R is 3.6.1, meaning there have been 3 major releases of R, 6 slightly minor releases, and 1 minor fix/release for the current version.

1.5 How This Book Can be Used in a Class

In addition to aiding the applied researcher in individual tasks and contexts, we also envision this book being used as a complementary text in the classroom. Whether in substantive social science classes that include a computational component, or in computationally intensive classes within the social sciences, we encourage widespread use of this book for the purpose of developing more efficient coding and programming practices.

To assist in classroom use, we design each chapter as a fully self-contained R session. This means that there are no parts of the online code sets that are left unexplained, no need to refer to earlier chapters to check whether a step is missed, no conflicts with earlier code, and no concerns about losing information for later if you shut down your R session. This does result in a certain amount of repetition in the code, which is a strategic choice on our part. As the Russian proverb states, "Repetition is the mother of learning." Similarly, there is strong scientific evidence that timed repetition is critical for mastery of tasks like mathematics, foreign languages, and computer programming (Oakley, 2014). For certain steps that are a part of any R session – setting a working directory, loading data, loading libraries – we want them to become automatic for the reader. Once this book is finished, this repetition should allow the user to immediately begin their given task without having to remind themselves of basic steps.

We also embed practice questions throughout the chapters for readers to work through and keep in mind as they code. Given our goal of efficient programming and deeper understanding of rigorous technical process, we will also include many conceptual high-level questions in these exercises. These "substantive pauses" in chapters reinforce our main goal in this book, rather than providing just a technical manual with a list of useful functions. Exercises are usually divided into "basic", "intermediate", and "advanced." The basic problems usually ask for modifications of code already introduced and require minimal understanding of what is happening and/or the relationship between concepts. These are for those who just want to do the things introduced in the book quickly. The intermediate problems will usually ask for some deeper understanding of what is happening in the introduced processes and may ask the reader to find new information. Finally, the advanced problems ask readers to undertake either a more complex task or to find some information on a related process not formally introduced in the chapter. Completing these exercises should result in the reader becoming comfortable with undertaking unfamiliar tasks on their own.

Finally, we provide a range of online documentation to assist in classroom instruction. Many of the most difficult challenges that instructors face are completely ignored by introductory textbooks, usually because they are considered "too basic" for inclusion. We provide things like step-by-step installation instructions for R and RStudio, additional tutorials on basic statistics, and script versions of all the chapters that can be downloaded, directly updated (if so desired), and used in classes. This collection is ever expanding, so instructors can check to see if we have a resource for their particular needs (and contact the authors with requests for materials that are not yet covered). All of these resources are free to use or modify for the instructor's needs, and, in the same open spirit, we encourage direct communication with students, researchers and instructors. The book website is at https://i2rss.weebly.com.

Instructors will probably choose to emphasize some parts of the text over others. For example, undergraduates may not need the introduction to programming that is included later in the book, and this can be safely skipped. Similarly, institutions that have a separate data visualization course may decide to focus elsewhere and have students use a more specialized text to learn visualization. We have tried to make each chapter relatively self-contained so that instructors can pick and choose if that fits with what they are trying to accomplish in their courses.

1.6   Plan for the Book

Here is the plan for the rest of the book.

Foundations: Chapter 2 provides the building blocks for the rest of the book. It starts with using R as an interactive environment. It then discusses the foundations of R – objects and functions. This sets up everything that follows, and will allow readers to understand how R is functioning throughout. Next, we introduce the process of setting working directories and working with R packages, both necessary for any work with R. Finally, we introduce the reader to the packages used in the book and where they can go for extra help.

Data Management: More than 80% of data analysis is data management. This chapter provides details of how to conduct most major data management tasks using a tidy approach. This includes selecting variables, filtering data, summarizing data, conducting summaries by groups, combining commands with the pipe (`%>%`), reshaping data, and combining data sets. Within this context, the chapter will also introduce how to create cross-tabulations and comparisons of means, since these types of analysis are a natural extension of data management. This chapter also introduces the `stargazer` package to automatically generate publication-quality tables in R (Hlavac, 2018). We have found this to be an entry point that is easy to understand for beginners, building confidence in their use of R throughout the rest of the learning process.

This chapter also serves as a practical introduction to the philosophy and structure of the Tidyverse.

Visualization: The next chapter deals with visualization and graphics in R, introducing the `ggplot2` package (Wickham et al., 2019b). This chapter focuses on the structure of graphics objects using the "grammar of graphics," which is at the heart of `ggplot` (hence, "gg"). It also introduces concepts like the scope of a variable through examples in graphing, mapping aesthetics, and layering plot objects. The chapter does not go into nearly the detail of other books focused on graphics, but it provides a general entry point to creating basic and advanced graphics in a tidy manner.

Programming: While base R is only used sparingly throughout the rest of the book, this chapter gets into much greater depth. We introduce the data structures of R (vectors, matrices, lists, data.frames), as well as the attributes of basic classes (character, numeric, factor). Students learn about indexing for each data structure, as well as some of their unique behaviors. This leads into a discussion of how to use conditions and loops to automate repetitive tasks, and how to save those programs as user-defined functions. We end by discussing the creation of modular code and providing some examples of useful functions.

Exploratory Data Analysis: One of the first steps in any data analysis is getting to know your data through exploratory data analysis (EDA). In this chapter, we begin discussion of statistical analysis in R by introducing some of the tools, especially the `skimr` package, for conducting this type of analysis (McNamara et al., 2019). We demonstrate how to visually and numerically analyze data using R, as well as how to "skim" the data to provide powerful extensions beyond the traditional `summary()` command in base R.

Essential Statistical Modeling: The final substantive chapter takes the reader through an introduction to correlation and regression in R, demonstrating how to conduct some of the most common types of analysis in the social sciences. We demonstrate t-tests, chi-squared tests, and regression, as well as a range of diagnostic tests, all based in the tidy R approach. For example, we use the `broom` package (Robinson, 2014) for tidy inspection of model output, and the tidy-friendly `performance` (Lüdecke et al., 2019) and `see` (Lüdecke et al., 2020) packages from *easystats* to diagnose and visualize influential observations. We also introduce logistic and probit regression and offer a demonstration of how they differ from OLS regression in R. We also show the reader how to automatically create publication-quality tables of their regression models.

Finally, we will conclude with a few thoughts on where the reader can go from here, as well as some parting tips for making the most of your R analysis.

2

Foundations

The goal of this chapter is to introduce you to key concepts used throughout the book. The first part will focus on philosophy and terminology of R. The second part of this chapter will focus on setting up the packages and libraries you need and how to download useful additional tools. The third part will introduce some of the resources you can use to help you out as you develop your R skills.

We will not be going into any kind of depth about the underlying design philosophy of R or some of the deeper programming principles of Base R. There are plenty of other resources for readers to obtain this kind of knowledge (Matloff, 2011; Leemis, 2016). Our goal here is to construct a solid platform for you to conduct a wide range of social science work. After reading this chapter, you should find it relatively easy to follow the subsequent chapters, as well as to utilize online tutorials. We are assuming absolutely no previous experience with R and RStudio. For those of you who have some previous background with using R and RStudio, you can safely skim through some of the parts that might already be familiar, but we do recommend at least taking a casual glance at the subsections of this chapter to make sure we are not leveraging something with which you may not have previously worked.

Before you start this chapter, be sure you have R and RStudio downloaded and installed, since we will be using both. If you need help with this process, or would like to start by having a more detailed understanding of the various parts of RStudio, you can consult with the online resources before moving on.

2.1 Scripting with R

When you open RStudio for the first time, it will look like the picture in Figure 2.1. You will see that it splits your screen into three parts. On the left, you will see a window that has the R console. Think of this like the command line for R. Inside the window is a > which shows you that R is ready to accept a command. When you type something into this window and hit enter, R will execute (or "run") that command.

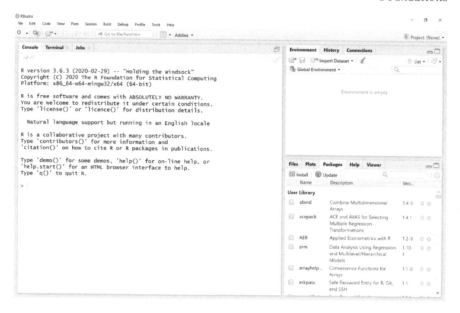

FIGURE 2.1

A View of RStudio Opened for the First Time

Give it a try. Type 1 + 1 in the console and hit Enter.[1] You will see that R prints the result, 2, on the screen.

1 + 1

[1] 2

This interaction within the console is one of the key features of R. R is what computer scientists call a "scripting" language. This means that, when a command is passed to R, it is evaluated immediately. In contrast, "compiled" languages, like C or C++, are ones where there is an intermediate step between writing the command and running the command. The code is compiled into native machine language before being run. While programs in compiled languages tend to run faster, the advantage of scripting languages is that they are usually easier to learn and allow for closer interaction in the context of data analysis. Moreover, as computers have become more powerful, the speed advantages of compiled languages (except for very intensive tasks) has tended to dissipate.[2]

---

[1] Note that you can either include or omit spaces between values. While the command will run either way, we recommend spaces between values in a function call to allow for easier reading of the code. E.g., 1 + 1 is cleaner and easier to read than 1+1; this will become clearer as functions and commands get more complicated later in the book.

[2] Note that recent efforts have been made to combine the efficiency and speed of scripting and compiled languages. For example, there are R packages that will allow you to interact with C++, Java, Python and other languages and tools. Indeed, many of the functions in base R are actually written in C!

Another feature of R that many users notice immediately is the ubiquity of parentheses. R uses parentheses to show where commands begin and end. A common problem users will experience is forgetting to close the parentheses on a command. If you do this, the > on the left will turn into a +, indicating that R is expecting more from the command. It will not return to the > until you have finished the command you are on. To demonstrate this to yourself, try typing (1 + 1, *without closing the parentheses.* Then, hit Enter/Return as many times as you would like. You will notice it will not give you the answer until you close the parentheses.

```
(1 + 1
```

```
)
```

```
## [1] 2
```

On the right-hand side of our screen when you first open RStudio, you will see two windows. On the upper-right-hand side, you will see a series of tabs, labeled "Environment", "History", "Connections", and (sometimes) "Build." We will only be using two of those tabs in this book. The Environment tab shows all the objects you have stored in memory for use in your R session. So, for example, if you load a data set, you will see it show up in the Environment tab. The History tab records all the commands you have made recently. If you ever want to enter one of these commands into the Console, you can simply double-click on it.

On the lower-right-hand side, you will see five tabs. The "Files" tab shows all the files in your working directory. You can use this interactively to see what you have in your working directory, as well as any other areas of your file system. The "Plots" tab will show you any graphs that you make, and allow you to export them for use in your publications. We will use this extensively in our chapter on producing plots. The "Packages" tab shows you all the packages you have installed, and will check any that you have loaded into your environment (more on this below). The "Help" tab can be used whenever you call for the R documentation on a function. For example, if you type help("cor") or ?cor, it will show you the R documentation for the cor() function (which, as you might guess, calculates the correlation between variables). The Viewer tab is for viewing local web content. We will be using this when we design three-dimensional plots later in the book.

Once you have opened RStudio, you can open the script containing your code for an analysis by either selecting File » Open File in the dropdown menus, or by clicking on the folder icon. You can create a new script by selecting "File" » "New File" » "R Script" from the dropdown menu or by clicking on the blank page icon and choosing R Script. You will also notice that beside the option to open a new R Script, it says "Ctrl+Shift+N." This is a keyboard shortcut. It means that if you want to open a new R script, you can do so by holding down the Ctrl key, the Shift key, and the N key at the same time (your keyboard shortcuts might be different, depending on the operating system you are using).

When you open an existing script or create a new one, a new box will appear above the console with the script. If you open more than one script at a time, you will see that it creates new tabs for each additional script.

We have provided scripts containing all the code from this book on the companion website.

Once you open a script with R commands in it, you will want to send those commands to the R console to run. You can do this either by clicking the Run button in the upper-left-hand corner of your script or by using the keyboard shortcut Ctrl+Enter (this may be different on your operating system; hover over the "Run" button to see how it is labeled on your system).

By default, RStudio will run only the line where the cursor is located. If you wish to run more than one line, you can highlight all the lines you wish to run.

Take some time to explore RStudio before you proceed to make sure you are comfortable with its operation. A more detailed version of the instructions above, with illustrations, is available on the companion website.

Exercises

2.1.0.0.1   Easy

- Practice interacting with the console. What happens when you type `"Hello World"` into the console and press Enter? What happens when you type `3492 / 12` and press Enter?
- Open a new blank script in three ways: by going to File » New File » R Script, using the "Ctrl+Shift+N" keyboard shortcut, and clicking on the new script icon. In one of these scripts, type the commands from questions 1 and 2. Run them from the script file.
- To paste ("concatenate") together more than one string, you can use the `paste()` function. Try this. Type in `paste("Hello", "World!")`. What happens when you put in a number, like `paste("Hole in", 1, "!")`?

2.1.0.0.2   Intermediate

- What happens when you type the following command into the command line, `round(sqrt(122.563^2, 2)`? How would you correct this?
- What error message do you get when you type in `"two" + 2` (be sure to include the quotation marks)? What do you think this means?

- In problem #3 above you might have noticed that `paste("Hole in", 1, "!")` produced a sentence with a space between the "1" and the "!". Why do you think this happened?

2.1.0.0.3 Advanced

- Previewing what we will see below, type `"two" * 2` into the console. What error message do you receive? What do you think this means? Feel free to look up the error message online to help.
- We will not discuss all the possible mathematical operators for R in this book, but there are a number of additional operators about which you might be interested. What do `**`, `%%`, and `%/%` do? You can look for information online.

2.2 Understanding R

The foundations of R are pretty simple, but are often a stumbling block for new users. Two general rules that we will often return to are:

1. Everything in R is an object.
2. Anything that does something is a function.

2.2.1 Objects

Just like objects in the real world, objects in R have "attributes." For example, the number 1 and the string "one" are both objects in R, but they have different attributes. These attributes determine what you can do with them. For example, adding two numbers makes sense, so running `1 + 1` in the R console will produce an outcome. Adding two strings does not make sense, so running `"one" + "one"` in the R console will return an error message.[3]

We will often use the `class()` function to get some information about the objects. For example, if you run `class(1)` in the console, it will return `numeric`. Alternatively, if you run `class("1")`, it will return that it is a `character`.

```
class(1)
```

```
## [1] "numeric"
```

---

[3]Some programming languages, like Python, will allow you to use mathematical operators with non-numeric values. For example, in Python, `"one" + "one"` would produce "oneone" (the same outcome as `paste("one", "one")` in R) or `"one" * 3` would produce "oneoneone" (the same as `rep("one", 3)` in R).

```
class("1")
```

```
## [1] "character"
```

There are times when we will simply want the objects printed in the console, and other times when we will want to save those objects for later use. To save an object to memory in your R session, you can use assignment operators, which are either <- or =. When you do this, it will appear in the Environment tab in the upper-right-hand corner of RStudio. The <- and = are synonymous, but most R users, by convention, use <- for object assignment.[4] In RStudio, you can also use the keyboard shortcut "Alt + -" to create <- ("Option + -" in Mac OS).

So, in the example that follows, the first line will simply print the results of 1 + 1 to the console. The second saves the object to memory and calls it two. To print this in the console, we simply run two in the console and it prints the object on the screen.

```
1 + 1
```

```
## [1] 2
```
```
two <- 1 + 1
two
```

```
## [1] 2
```

R is what is called a **strongly typed** language. This means that capitalization and punctuation are important. The object Two is not the same as the object two. Getting an error returned saying that an object does not exist is often due to spelling or capitalization mistakes. Here is a quick example.

```
two <- 2
Two <- 2.2
two
```

```
## [1] 2
```
```
Two
```

```
## [1] 2.2
```
```
two == Two
```

```
## [1] FALSE
```

---

[4]The original computers used by Bell Labs in creating R had a single key that produced this assignment operator. Most users still prefer it today both for style reasons and because of some rare situations where the = may demonstrate unexpected behavior.

2.2.2 Functions

Functions in R work the same way as the functions you learned about in elementary school math. They take an object, do something to it, and return another object. Functions in R are usually denoted by their use of parentheses. As we mentioned, everything that does something in R is a function, making R a "functional programming language."

As you might have guessed from the last paragraph, you have already learned a function in this book. The `class()` *function* takes an object as its input and returns the name of the class of the object as a character string.

The inputs into a function are called "arguments." And running the function is called a "function call." Some functions take a single argument and return a single object. Other functions can take on many arguments, and can return many objects. Some functions will have "default" arguments and behavior, so you do not need to type every input. For example, if you type `help()` into the console (with *no* argument), the Help tab in the lower-right side of RStudio will bring up the documentation for the `help()` function. If you type `help("lm")`, you have passed an argument to it (labeled "topic" in the `help()` documentation) asking for the help documentation on fitting a linear regression model (`lm`), and returns this documentation instead of the default.

So, you might be wondering how functions fit within the rule that everything is an object in R...

Functions are *also* objects.

For example, if we look at the class of the `help()` function, we will see that it returns an object class – in this case it is a special class `help_files_with_topic`.

2.2.3 Commenting with

Comments are invaluable when working in R. Comments allow for a variety of tasks including saving some code for later, altering a chunk in the script editor, or including notes to yourself to reference in future runs of code in an R script.

To include a comment or to "comment-out" a chunk of code, simply place # before the text or code to be commented which equates to *not* running whatever may follow the #.

For example, suppose you wanted to use the Pythagorean Theorem ($a^2 + b^2 = c^2$) to solve for the hypotenuse, c, of a triangle. To solve for c, you take the square root of the sum of the other two sides of the triangle, $c = \sqrt{a^2 + b^2}$. You could build a function to solve for this based on any two values supplied for sides a and b by starting with the following:

```
solve_for_c <- function(a, b){
  sqrt(a^2 + b^2)
}
```

Though this function technically does the job, suppose you wanted to follow best practices of function-writing, and wanted to make this function *defensive*.[5] To do so, you might be interested in testing out various `warning` messages based on a mistake on the part of the user. So you might update the function to be:

```
solve_for_c <- function(a, b){
  sqrt(a^2 + b^2)
  if (!is.numeric(a)) {
    stop('"a" must be numeric\n',
         'You have provided an object of class: ', class(a)[1])
  }
}
```

This warning would let the user know 1) whether a non-numeric value was supplied to the function, and 2) the class of the object(s) supplied. Though these ideas and terms are covered in depth later in the book, the point here is that you could use the # to comment-out the warning message and redefine the function to test it in real-time, e.g.,

```
solve_for_c <- function(a, b){
  sqrt(a^2 + b^2)
  #if (!is.numeric(a)) {
  #  stop('"a" must be numeric\n',
  #       'You have provided an object of class: ', class(a)[1])
  #}
}
```

When you call your function `solve_for_c()` with the warning commented out, the call will ignore all of the code following each # and simply run the calculation included at the outset in the simplest version of the function.

A few things to note when using the #. First, when commenting multiple lines, you must include the # before *each* line; otherwise, calling the function will throw an error. Second, especially in larger chunks of code, be sure that only the parts you intend to comment-out are indeed commented out. The most common error in this regard is commenting out a chunk of code as in the example above, but forgetting to also comment out the closing } on the penultimate line of the function. Here again, failing to do so and then attempting to call the function will throw an error. And finally, we recommend

---

[5]Note: we cover defensive programming in the context of user-defined functions in the *Essential Programming* chapter.

liberal use of comments, especially when writing large chunks of code or writing code in real-time (e.g., during class or a lab session). Comments in this context are exceedingly valuable for annotating complex code that you may forget when you return to the script. You will see us including comments in code throughout this book.

Exercises

2.2.3.0.1 Easy

- In the previous exercises we used the `paste()` function to paste (concatenate) words and symbols together. You may have noticed that it automatically added a space between the items we are pasting together. For example `paste("Hole in", 1, "!")` places a space between "1" and "!". Type in `?paste()`.
  - Reading the help, what argument sets this?
  - What is its default?
  - How can you eliminate the space (or add something different)?

2.2.3.0.2 Intermediate

- Technically, in question 9, there are two ways to eliminate the space between words. What is the other way to do this?

2.2.3.0.3 Advanced

- In this section, we said that functions are usually denoted by parentheses, but you have seen symbols that do things without parentheses. Specifically mathematical operators, like +, -, /, and *. Type `?'+'` into the console. What does it tell you about these operators?
- Try typing `'+'(1, 1)` into the console. What does it produce? How does it show that these operators are, in fact, similar to other functions?

2.3 Working Directories

One of the first obstacles new users of R often face is understanding and setting a "working directory." The working directory is the place on your computer from which you want R to work. Setting a working directory ensures you know from where you are navigating to find files and to where the objects you want to save are being placed.

Whenever R is opened, a working directory is automatically assigned. To see where this is, users simply need to run the function `getwd()`, with no argument in the parentheses. This will return a single line in the output console with a file path for the location of the currently assigned working directory. If users are happy with this location, no further action is needed.

But if you want to change this, then they need a slightly different command, setwd(), with the name of the new file path included in quotation marks in the parentheses. For example, typing the following, setwd("/Users/username/Desktop"), will set the working directory to be on the desktop of a Macintosh computer for user, username. For Windows users, the command will take the form setwd("C:/Users/username/Desktop/").

While we recommend users become familiar with their file system and set their working directory by command, those who are less familiar with their computer's file system may also set their working directory using an interactive browser. This can be accessed through the dropdown menu by going to Session » Set Working Directory » Choose Directory (or using the "Ctrl+Shift+H" keyboard shortcut). In the examples in the following chapters, we will use the command setwd(choose.dir()), which also allows the user to interactively set the working directory if using Windows or Mac OSX, but we strongly recommend you become used to setting your working directory using one of the other methods, or that you start using R projects, which are explained in the next section.

2.4   Setting Up an R Project

In the previous section, we explained how to set up your working directory using the setwd() command. While setting working directories is a fundamental part of many computer programs, doing so often causes unnecessary problems. Effectively using the command requires that you understand how your directories are organized on your computer, and this tends to vary by whether your operating system is Windows, Linux, or OSX. For example, in the last section, we showed how to set the working directory to the Desktop folder on a Mac using the command setwd("/Users/username/Desktop"). But, for a Windows user, this command will not work, and will generate an error, Error in setwd("/Users/username/Desktop") : cannot change working directory. This is because Windows starts in a different area of the computer – the C drive. To do the same in Windows, the command becomes setwd("C:/Users/username/Desktop").

If you use more than one computer for your work, this can also cause issues. If, for example, your computer at work has the username rkennedy, but your laptop has the username Ryan, you will have to change your directory every time you try to use your code.

Things get a lot worse when you start working with other people. Their directories are likely to be structured differently than yours, meaning that they cannot just run your code – they must find where you make reference to a directory and change it. This can get so annoying that one Twitter post to the #rstats discussion thread threatened to set a user's computer on fire if their code included the `setwd()` command.[6]

While threatening to set someone's computer on fire may be a little extreme, the reality is that including information that will only work on your computer is inconvenient for you (at least if you ever plan on doing work on a different computer) and discourteous to anyone with whom you work.

Luckily, there is another way to do things. In RStudio, you can set up a "project." The project stores information needed to run your code and find your files, without you always having to tell it where to look. It also makes sure that those with whom you are working do not revolt when working with your code.

To create a project, just go to the Project menu in the upper-right-hand corner of RStudio, and select `New Project`. Once you have done this, you will be asked if you would like to create the project in a new directory or an existing directory. If you already have a folder containing your data, you might choose an existing folder. If you are starting from scratch, or simply want a new folder with which to do your work, choose a new directory. Locate the area into which you want to put your project, and, if it is a new directory, give it a name. For step-by-step instructions with illustrations, you can go to the book's support website.

RStudio will create a new file with a `.Rproj` extension. Whenever you open this file, either by double-clicking on it or navigating to it using the Project menu, it will automatically set your working directory to the location of the `.Rproj` file. If you copy the folder to a new computer - no problem, all your code will still work. If you work with a co-author through Dropbox or another shared system - no problem, they can simply open the `.Rproj` folder and it will work. (*Note*: If using Dropbox, you may want to pause syncing while you are working on a project to avoid error messages.)

In subsequent chapters, we will use the `setwd(choose.dir())` command to set the working directory so that each chapter is self-contained, but in some of the online scripts we provide examples of `.Rproj` files as well, and we recommend you get used to using and creating these.

---

[6]https://www.tidyverse.org/articles/2017/12/workflow-vs-script/

2.5 Loading and Using Packages and Libraries

Packages are a fundamental part of R. Packages contain many useful libraries of functions that you will need for your work in R. In 2017, the Comprehensive R Archive Network (CRAN) surpassed the 10,000 packages mark and it is still growing. R users worldwide find solutions to the data and analysis challenges they face, and they share these solutions on CRAN, GitHub, Zenodo, and other sites.

This aspect of working in R is incredibly valuable for a couple of reasons. First, for almost any task you can think of, someone has likely written a package to make your life easier. This is one of the main advantages to R being open source. Some of these packages you will use all the time – indeed, they may become a default heading on your code. Others you may just use for one project.

R often has capabilities years before other statistical programs. In fact, a common reason why people learn R is that they find out that a particular task has already been developed as a package in R, while it has not been implemented in the statistical package they usually use.

Second, the use of packages means that you only have to install what you need for your current project. This can seem a little odd for those used to working in other statistical programs that automatically load everything the program can do every time it is opened. The reality is that R simply has too much that it can do to load everything every time (remember, there are more than 10,000 packages available, containing millions of functions). By only loading what you need, you ensure that your projects are only using the resources that you actually need.

2.5.1 Installing Packages

To access packages, users must first *install* the package, and then *load* the library to be able to use the functions stored within the package source files. The basic function for installing any package is `install.packages("PACKAGE_NAME")`.

For example, to install all of the Tidyverse packages that are used throughout this book, users would install the `tidyverse` package (yes, that includes many packages in a package):

```
install.packages("tidyverse")
```

RStudio provides an additional way to do this. In the lower-right hand window, in the "Packages" tab, you can click the "Install" button. Once you have done this, you can type in the packages you want and click "Install" to install them. This can be useful because RStudio will list the packages that match your search as you type, avoiding common errors in spelling or capitalization.

Once you have installed a package, it is on your computer and you do not need to install it again when you restart R or start a new project, though you may need to reinstall packages when you update your R and/or RStudio versions.

2.5.2 Loading Packages

Once you have the package installed, you need to load the library of functions into your workspace using the command `library(PACKAGE_NAME)`. Note the quotation marks in the `install.packages()` command, but the lack of quotation marks in the `library()` command. You will run up against error messages if you reverse these.

Here is an example of how to load the `tidyverse` package, now that it is installed (per the above line):

```
library(tidyverse)
```

To use a package, you will need to load it using the `library()` function every time you restart R or start a new project.[7]

Now that you know the general use of packages and libraries, we will provide an overview of some of the main packages we will use throughout the book.

2.5.3 The `here` Package

An example of a very simple package we will be using throughout the book is the `here` package. When you are navigating through your file system, it is usually done using a string like `Users/username/Desktop`. These can sometimes get very long and annoying to type, especially if the file we are looking for is deep in our file system.

The `here` package is incredibly simple, but also incredibly powerful, especially when combined with an R project. After installing the package, you can load it into the workspace using `library(here)`, and you will see it print out a message that reads, `here() starts at <DIRECTORY>`, where the directory is the current working directory.

```
library(here)
```

The `here` package allows you to use the `here()` function to create a string with the information you need to point R to a particular file. For example, if

---

[7] *Note:* alternatively, you can save libraries to your R profile, but we do not recommend doing this because your needs will likely change over time.

we have set our working directory to the desktop and we want R to access the `dataset.csv` file in the `Data` folder on the desktop, would simply use `here("Data", "dataset.csv")`. As you can see in the code block that follows, this simply outputs a string with that location.

```
here("Data", "dataset.csv")
```

```
## [1] "/Users/waggoner/Dropbox/.../Data/dataset.csv"
```

Whenever we use a function to load our data, we can place this `here()` call into the function to load the data.

```
read_csv(here("Data", "dataset.csv"))
```

The beauty of the `here()` function is that if we use a different computer, with a different file structure, we can use the same exact command on both computers. This is very nice when you are working on multiple computers or with collaborators on a project. We will be using it in this book so that the code we provide will work on your computer without modification.

2.5.4 The `tidyverse` Package

Another package we will be using regularly in this book is the `tidyverse` package. As noted above, the Tidyverse is actually a set of packages that share a common philosophy and grammar. This includes the `ggplot2` package for creating graphics (used extensively in the *Visualization* chapter), the `tibble` package for producing a tibble data structure that has some useful properties compared to R's default `data.frames`, the `dplyr` package that allows for quick and readable data manipulation, the `tidyr` package to reshape your data into a "tidy" format that is useful for analysis, the `readr` package for quickly parsing a range of "rectangular" data structures that are common in the social sciences (these four packages are introduced and used extensively in the *Data Management* chapter), and the `purrr` package for functional programming that makes some repetitive tasks much simpler.

You could install and load all of these packages individually, but since these packages are commonly used together, they have been bundled in a single `tidyverse` package to make it easier to load.

While we strongly advocate the Tidyverse approach to programming and working in R, it is still useful to understand some parts of base R. You can do just about any basic data analysis task in the Tidyverse, but there may be some situations that require you to program in base R, you may see an example that uses base R, or you may encounter a package you want to use that follows base R conventions. Thus, we will introduce some base R throughout the book.

2.5.5   Overlapping Functions

In some cases, different packages will use the same name to do different things. For example, in the next chapter, we will be using the `dplyr` package that is included in the `tidyverse`, and we will be making use of one of its component functions, `select()`, which selects named columns from a data set.

Perhaps unsurprisingly, there are other packages that also have `select()` functions. One of them is the `MASS` package and another is the `skimr` package. If you load all of these packages and try using the `select()` function from `dplyr`, the system will not know which one to use. If this occurs (usually indicated by an "unused argument" error), you need to specify which package's function you want to use, separating the package and function name with `::`. For example, by running `dplyr::select()`, you are stating "use the `select()` function from (`::`) the `dplyr` package.

2.5.6   Other Packages Used in This Book

There are a number of other packages we will be using in this book. Here is a list of all of them, along with a short explanation of what they do.

`here` is a package that allows you to interactively search your working directory. For example, if you want to navigate to the "Data" folder in your working directory and find the "raw_data.csv" file, you would use `here("Data", "raw_data.csv")`.

`readxl` is a package for opening Excel spreadsheets (.xls or .xlsx) in R.

`haven` has functions for opening Stata (.dta), SPSS (.sav), and SAS (.sas7bdat) data.

`stargazer` is a package for producing professional tables that can be imported into other common word processing software like Microsoft Word or LaTeX.

`forcats` is a package that provides a set of functions for handling "factor" variables.

`corrr` provides functions for analyzing the correlation between variables that is more detailed and intuitive than base R's traditional `cor()` function.

`janitor` is a package to make nice-looking cross-tabulations, with many options for customization and calculation.

`purrr` is a Tidyverse package housing, among many other useful functional programming tools, the map family of functions covered in the *Essential Programming* chapter. Mapping functions are important Tidyverse innovations, allowing social scientists a streamlined bypass of `for` loops and the base R `apply` family of functions.

`amerika` supplies a color scheme that mimics traditional colors for graphics about American politics (i.e., red for Republicans and blue for Democrats) (Waggoner, 2019).

`arm` is a compilation of many useful packages for analysis associated with Andrew Gelman and Jennifer Hill's popular book, "Data Analysis Using Regression and Multilevel/Hierarchical Models" (Gelman and Hill, 2006).

`faraway` is a compilation of data sets and functions from Julian Faraway's book, *Extending the Linear Model with R: Generalized Linear, Mixed Effects and Nonparametric Regression Models* (Faraway, 2016).

`MASS` is a compilation of data sets and functions from Bill Venables and Brian Ripley's book, *Modern Applied Statistics with S-PLUS* (Venables and Ripley, 2013).

`OOmisc` contains a set of useful miscellaneous functions produced by Ozgur Asar and Ozlem Ilk.

`pROC` is a package of functions to produce and analyze receiver operating characteristic (ROC) curves.

`lmtest` is a package of functions for analyzing regression models, including likelihood-ratio tests.

`rstatix` is a package for evaluating basic statistical functions such as t-tests.

`car` provides functions utilized in John Fox and Sanford Weisberg's book, *An R Companion to Applied Regression* (Fox and Weisberg, 2018).

`plotly` is a powerful package for advanced plotting, including interactive plots.

`broom` is a Tidyverse-complementary package for inspection of model objects, which is much more thorough than the `summary()` function in base R.

`patchwork` is a Tidyverse-complementary package for placing `ggplot` objects in a single pane with minimal code.

`performance` is a package from the *easystats* software group that includes a host of performance checks for regression models.

`see` is a Tidyverse-complementary visualization package from the *easystats* software group that complements `ggplot2`, and also allows for plotting objects created using the `performance` package.

You can install all of the packages needed for this book by running the following code chunk in your console. Alternatively, you can click on the "Packages" tab in the lower-right-hand corner of RStudio, click "Install," type in all the packages you want (separated with a comma or space), and click "Install."

```
install.packages(c("tidyverse", "here", "readxl", "haven",
                   "janitor", "stargazer", "forcats", "skimr",
```

```
              "corrr", "amerika", "purrr", "arm",
              "faraway", "MASS", "OOmisc", "pROC",
              "lmtest", "car", "rstatix", "plotly",
              "broom", "patchwork", "performance", "see")
      )
```

Exercises

2.5.6.0.1 Easy

- Try installing and loading the `arm` package from Gelman and Hill's book (Gelman and Hill, 2006). Make sure you understand this process.
- Install the other packages needed for this book. Try loading `tidyverse` and `here`.

2.5.6.0.2 Intermediate

- You may have noticed that we used the `c()` function to create a vector of packages we wanted to install. Run `?c()`. What does this tell you about the `c()` function? What happens when you type `c(1, 2, 3)` into the console? Why?
- Set up an R project called "R Code" somewhere on your system. You can either create the folder and then create the project, or you can create the folder by creating a project. What happens when you open the project? Now create subfolders in the project location for "Data" and "Code". Use the `here()` function to create a string that indicates these subfolders.

2.5.6.0.3 Advanced

- Some even newer packages are available on GitHub, a repository for programs and packages that is open to anyone. How would you install a package from GitHub? Try installing the package `fliptime`, which might be useful for those of you working with data with calendar dates, from the GitHub address `"Displayr/flipTime"`.

2.6 Where to Get Help

We all need help every now and again. Even after more than a two and a half combined decades of using R between the authors, we still often seek help with how to do some tasks. The reason is not so much that R is difficult to use, but rather that, when you try to do something new in any system, you will run into unfamiliar challenges. Even Hadley Wickham, the designer of the Tidyverse, is there with you (Waggoner, 2018a)...

It's easy when you start out programming to get really frustrated and think, "Oh it's me, I'm really stupid," or, "I'm not made out to program." But, that is absolutely not the case. Everyone gets frustrated. I still get frustrated occasionally when writing R code. It's just a natural part of programming. So, it happens to everyone and gets less and less over time. Don't blame yourself. Just take a break, do something fun, and then come back and try again later.

This is where R's amazing user community comes in. Our experience is that, because of R's large user base of people who are trying all sorts of creative projects, it is usually easier to find help with the challenges that arise in R than it is in other statistics packages. There are several ways to find help.

If you just need a reminder of how a function works or what options are available for a command, you can use R's official documentation. Within RStudio, as we previously noted, this can be accessed simply by typing a ? followed by the command about which you have a question, or by typing help() with the command you would like help with in the parentheses in quotation marks. For example, to quickly remind yourself of the arguments and options available for regression models, you could type either ?lm or help("lm") and the official documentation for the function will show up in the help tab in the lower-right-hand corner.

Beyond the official documentation, there is a vibrant R users community that is very willing to help people learn and deal with new situations. As the Beatles would say, "I get by with a little help from my friends." In R, you have tens of thousands of friends willing to help you out. A good place to start is R-Bloggers or the "#rstats" hashtag on Twitter (https://twitter.com/hashtag/rstats). R-bloggers compiles blog posts from a range of authors. It is a great place to find announcements about new R packages and books, available courses, tutorials for different tasks, and just about anything else you can do with R. You can subscribe to receive daily emails that are usually filled with interesting tidbits about what you can do with R.

"Coding by search engine" is really a thing. We have heard tenured computer science faculty at highly prestigious institutions who have described their process using this phrase. If you want to learn how to do something new or if you get an error message, going to your preferred search engine and typing it in is usually not a bad place to start. For example, typing "regression analysis in R" into Google, will produce a number of tutorials to take you through examples. Often your searches will be more effective if you know the package you want to use. So, for example, we often use "tidyverse" or "dplyr" in our searches for data management questions. Similarly, if you get an error message and copy it into a search engine, chances are you will be directed to a site where someone has posted the question and someone else has answered it. We have almost always found that when we encounter an issue, someone else has also encountered it and has posted a solution. The largest repository of solutions posted by R users just like you is Stack Overflow (https://stackoverflow.com/).

The main thing to remember is that you are not alone. If you are running into a problem, chances are that there are many others out there who have had the same problem, and, because of this, have probably put the information you need in the help files or posted a solution online. Nothing worth doing is going to be without some frustrations, but there are plenty of places to help you when you struggle. Now, let's move on to the first step in almost any data analysis – data management.

2.7 Concluding Remarks

If you haven't been working through the problems in this section or following along with the examples, we suggest you take some time to do so before you move on. The online resources will take you through some code examples and will also provide you with illustrations of some of the procedures in this chapter. Just as a building is only as good as its foundation, you will have difficulty proceeding if you do not understand what we have done so far. You do not need to have everything mastered, but you do need to have a basic idea of what we are talking about.

If you feel like we have been giving you a lot of information, do not worry. In subsequent chapters, as well as in the online resources, we will give you plenty of examples. Many of the things in this chapter, like setting a working directory, will be repeated in each chapter, both to make the chapters self-contained and to provide you with review. As numerous studies have demonstrated, learning skills like coding requires this type of repetition and reminders to promote mastery (Oakley, 2014). If you keep working through this, you will likely look back on this chapter in a few years and think it is too simple.

3

Data Management and Manipulation

One of the first tasks most users will encounter when they receive a new data set is to get the data in the form they want. This may involve a range of tasks such as reading data sets from different formats, combining multiple data sets, summarizing the data, creating new variables from the old variables, and a range of additional tasks. Data rarely comes in a form that is ready to use. There are often errors, improper formatting, missing data and a range of other issues researchers must address. As scholars have started leveraging less structured and more complex data, like "big data" and text data, the importance of data management and manipulation ("munging" or "wrangling" as it is sometimes called) has become even more important (Radford and Lazer, 2019). *InfoWorld* identified this as the 80/20 dilemma, where most data analysts spend 80% of their time in data management and manipulation, while spending 20% of their time in actual analysis.[1]

Yet, most introductions to computational statistical analysis might give a very short introduction to data manipulation. In some ways, this is not terribly surprising. We all got into the social sciences to make discoveries, not necessarily to spend our time shaping and cleaning data. Why can't we just skip to fitting statistical models that provide support for our research question?

We hope to convince you that good data management and manipulation is not only necessary, but can also be quite rewarding. Building strong skills in data management and manipulation will allow you to get to your answers faster, and will allow you to create data for answering novel questions. If all you ever learn is how to work with clean data created by others, you will only be able to answer questions addressed in others' data. This is not where you want to be as a social scientist.

This chapter will teach you the basics of managing your data, from loading the data into R to exporting the data to other programs and reporting the information about your data. It will cover all of these tasks using the Tidyverse, along with a few useful utilities in base R.

---

[1] https://www.infoworld.com/article/3228245/the-80-20-data-science-dilemma.html

3.1 Loading the Data

The first task is taking the data you have and reading it into R for analysis. In some programs, this can be an initial source of frustration. Statistical software programs usually want you to use their format for your data and may make it difficult to use data from another source (or even from different versions of the same program).[2] Because R has such a large user community, who also use data in a variety of formats, there are packages that will allow you to read data from a wide variety of sources.

Go ahead and open a new session of RStudio and set your working directory to the folder for your project. If you are working with R projects (the .Rproj files provided in the online code or a project you have created yourself), you can simply open the .Rproj file and it will automatically set your working directory to the location of the .Rproj file. You can then skip the following command. If you need a reminder about how R projects work, see the "Foundations" chapter.

```
# Set your working directory
setwd(choose.dir())
```

Let's start with one of the most common and simple formats. Comma-separated values (CSV, with the extension .csv) files are used for storing a lot of social science data, and is a standard format for programs like Microsoft Excel. This type of data stores everything as text, with commas separating the columns. The reason it is commonly used is that it does not require a lot of space to save and it works with almost any kind of statistical program. In addition, it never becomes out-of-date – while the ability to open particular types of data files may fall by the wayside with time, the ability to open and parse text files will not. To load a data set that is in .csv format, you can use the read_csv() function from the **readr** package that is included when you load tidyverse. We also use the here() function to tell the computer to look in the "data" subfolder of our working directory. If your data is already in your working directory, you can just use read_csv("anes_pilot_2016.csv").

```
# Load the libraries needed for this session.
library(tidyverse)
library(here)
```

---

[2] Programs that charge for upgrades are particularly notorious for changing their default data format. The goal is to force users to upgrade to the newest version. Even worse, sometimes they will not provide backward compatibility with previous data formats. It is always a good idea to keep at least one version of your data in a plain text format (comma-separated or tab-separated) so you will not find your data unreadable in the future.

```
# Load the NES data using the read_csv() and here() functions
NESdta <- read_csv(here("data", "anes_pilot_2016.csv"))
```

While CSV files are quite common, sometimes you will find data in a variety of other formats. For example, since Microsoft Excel is often used to store data, especially by people who work in business and public policy, you may also encounter data that is stored in its default format (.xls or .xlsx). Since other researchers have encountered this, there is an R package specifically for loading this kind of data called readxl. Here is how to load the same data if it were saved in this format.

```
# Load the additional library needed for this task.
library(readxl)
```

```
# Load the NES data using the read_excel() and here() functions
NESdata <- read_excel(here("data", "anes_pilot_2016.xlsx"))
```

Similarly, if we had a data set that had been saved using SAS (.sas7bdat), SPSS (.sav), or Stata (.dta) – three popular statistical packages in the social sciences – you can use the haven package.

```
# Load the additional library needed for this task.
library(haven)
```

```
# Use read_dta() and here() functions for Stata file
NESdata <- read_dta(here("data", "anes_pilot_2016.dta"))
```

```
# Use read_sav() and here() functions for SPSS file
NESdata <- read_sav(here("data", "anes_pilot_2016.sav"))
```

```
# Use read_sas() and here() functions for SAS file
NESdata <- read_sas(here("data", "anes_pilot_2016.sas7bdat"))
```

As we have already noted, one of the real powers of R is the ability for users to write their own solutions to address problems that are often encountered by researchers. In a number of other software programs, it can be quite difficult to read certain types of data unless you have access to that specific program. An (in)famous example of this occurred in 2013, when Stata changed its data format, such that older versions of the program could not open files saved by the new version of Stata. Anyone with older versions of the program found their system was rendered functionally illiterate overnight. R can handle all of these data types (plus many more) because its global user base has made it relatively easy to do so.

R also has the capacity to open files from database programs, like SQL, or from online APIs that usually report results in JSON format. We will not cover all of these here, but it is worthwhile to note that just about any kind of data structure can be handled by R with the use of packages.

3.1.1   The American National Election Survey (ANES)

For this chapter, and several of the subsequent chapters, we will be using the American National Election Survey (ANES) for demonstration. The ANES project is one of the longest running in political science in the U.S. While it was formally created by the U.S. National Science Foundation (NSF) in 1977, the University of Michigan had been conducting surveys around midterm and presidential elections going back to 1952.[3] The rich results from these surveys have been the raw material for countless books, dissertations, and published articles (Campbell et al., 1960; Lewis-Beck et al., 2008; Aldrich and McGraw, 2012).

We specifically look at the 2016 pilot study, which was collected between January 22 and January 28, 2016. 1,200 individuals were interviewed in a 32 minute online questionnaire. The survey included questions covering a range of topics among U.S. eligible voters: preferences in the presidential primary, stereotyping, the economy, discrimination, race and racial consciousness, police use of force, and numerous policy issues.

We choose this survey for several reasons. First, we want readers to see and practice with real data, warts and all. This is especially important for this chapter. Well-manicured data sets encourage spending more time learning specific statistical models, when a large portion of most researchers' time will be spent getting the data into the condition that is needed for the analysis. Second, we wanted to give researchers experience with data they might actually be interested in using later. Experience with this data set will allow the reader to work with other ANES data sets, which have many of the same characteristics. Updated ANES data is regularly becoming available and can be downloaded for use in research projects from the ANES website.[4] Third, the tasks demonstrated are similar to those readers will commonly see on the news, evaluating candidate support and popularity, giving a common reference point for readers from a variety of fields. Indeed, in just the first three months of 2020, there were 15 news articles about ANES data – more than one per week. Finally, it covers an interesting period of U.S. politics, e.g., a few days before Senator Ted Cruz and Secretary Hillary Clinton would win the Iowa Caucus, and a couple of weeks before Donald Trump and Senator Bernie Sanders would win the New Hampshire primary. While some of you may be reading this when all these events are a distant memory, this is a critical juncture in an election that would eventually lead to the election of Donald Trump as President, with very large consequences for U.S. domestic politics

---

[3]https://electionstudies.org/about-us/history/
[4]https://electionstudies.org/

(Pierson, 2017), the global political system (Giani and Méon, 2017; Ikenberry, 2017), and our understanding of how politicians win elections (MacWilliams, 2016; Sides et al., 2017).

3.1.2   A Short Note on Data Structures

In the examples above, we are loading the data using packages that are included by default in the Tidyverse or in packages that are designed in accordance with Tidyverse standards. Technically, this means that the files are being loaded as *tibbles* (class `tbl`).

```
# Check the class of the NESdta tibble.
class(NESdta)
```

```
## [1] "spec_tbl_df" "tbl_df"      "tbl"         "data.frame"
```

A tibble is one data structure in R, but it is not the only one.

The default data structure in base R is called a `data.frame`. A CSV file, for example, can be loaded using the `read.csv()` function as demonstrated below.

```
NESdata_df <- read.csv(here("data", "anes_pilot_2016.csv"))
```

Tibbles work in the same way as `data.frames`, but with some important modifications. In some ways they do less. For example, they will not automatically change your variable names or types. At the same time, tibbles also provide some nice additional options. For example, tibbles will complain more when there are issues with the data. While warning messages might be annoying, they can also be quite useful for flagging problems with data and helping to avoid problems down the line.

Tibbles also provide more information in an easier to understand format. If we try to print the NES data on our screen by typing `NESdata_df`, the results will be very long and almost impossible to read (indeed, we will not show it in the book because it is so messy). If we type in `NESdata` (our tibble), we get something much more comprehensible, providing only the first ten rows of the first few variables, along with the type of data, and then a printout of the rest of the variables available with their type.

```
# Show the structure of our NES tibble
NESdta
```

Some functions still require that you use the `data.frame` format. You can easily convert between tibbles and `data.frames` using `as_tibble()` or `as.data.frame()`, respectively. The following code chunk shows conversion from a data frame to a tibble.

```
# Convert data.frame to tibble
```

```
NESdata_df_tibble <- as_tibble(NESdata_df)
class(NESdata_df_tibble)
```

We can of course go the other direction from a tibble to a data frame.

```
# Convert tible back to data.frame
NESdata_df_tibble_df <- as.data.frame(NESdata_df_tibble)
class(NESdata_df_tibble_df)
```

Before we leave this discussion, a more general pattern is worth highlighting. You can see us using the "as" prefix to convert between different types of data. This is a more general pattern in R. If we want to convert a numeric data to factor variable, for example, we would use `as.factor()`. You will see several versions of this prefix as we continue through the book.

Exercises

3.1.2.0.1   Easy

- On the webpage for this book, we also provide a data set with information about the 50 U.S. states. How would you load the `states.csv` file and save it in memory as an object named `states`?
- Type in `NESdta` to see the tibble structure. What type of variables are there in this data set?
- Sometimes it is useful just to get a vector of names for variables. Use the `names()` function to get these for `NESdta`.

3.1.2.0.2   Intermediate

- We noted in the previous chapter that everything in R is an object and anything that does something is a function. How is the `here()` function an example of this? As a function, what does it do? As an object what attributes does it have? Run `class(here())`. What does this tell you?
- `class()` is also a function. Run `class(NESdta$fttrump)`. What does this tell you? How does the behavior of `class()` change?

3.1.2.0.3   Advanced

- We saw how to see the structure of a tibble. How would you do the same with a data.frame? [Note: You can look this up online.]
- The `names()` function gets the names of the variables in a data.frame or tibble. It can also be used to change the names of the variables. How would you do this?

3.2 Data Wrangling

With the data loaded, we need to start doing things with it. Data "munging" or "wrangling" is the process of getting your data into the form you need for analysis (i.e., data management). The Tidyverse offers a myriad of functions for effectively, efficiently and consistently managing data. Most of these functions are in the `dplyr` package, which is one of the main components of the Tidyverse. We will cover eight of these functions:

- `select()` - choose specific variables you wish to keep
- `filter()` - filter your data by selected values
- `group_by()`- group your data by categorical values
- `summarize()` - create summary statistics of data
- `join()` - merge different data sets
- `mutate()` - create new variables

As with almost any data set, the NES data has many more variables than we could ever really plan on using in a single analysis. So, we might want to limit ourselves to just those in which we have some interest. Let us create a new object, called `NESdta_short`, which includes only the variables we will need for this section. You will notice that throughout the book, we use `dplyr::select()` instead of just `select()`. As noted previously, several packages used in this book have their own versions of the `select()` function, so this ensures we are using the `select()` function associated with the `dplyr` package.

```
# Select particular variables
NESdta_short <- NESdta %>%
  dplyr::select(fttrump, pid3, birthyr,
               ftobama, state, gender, pid7)
NESdta_short
```

```
## # A tibble: 1,200 x 7
##    fttrump  pid3 birthyr ftobama state gender  pid7
##      <dbl> <dbl>   <dbl>   <dbl> <dbl>  <dbl> <dbl>
## 1        1     1    1960     100     6      1     1
## 2       28     3    1957      39    13      2     4
## 3      100     2    1963       1    24      1     6
## 4        0     1    1980      89    35      1     1
## 5       13     4    1974       1    27      1     5
## 6       61     3    1958       0    18      1     4
## 7        5     1    1978      73    23      1     1
## 8       85     2    1951       0    53      1     7
## 9       70     3    1973      12    18      1     4
## 10       5     1    1936      87    12      1     1
```

```
## # ... with 1,190 more rows
```

This creates a new tibble that only includes five variables: `fttrump` - a "feeling thermometer" where people rate their feelings of then primary candidate Donald Trump from 0 to 100; `pid3` - a three point rating of political identity, where 1 means Democrat, 2 means Independent, and 3 means Republican; the respondent's `year` of birth, which we will use to establish their age; the respondent's `gender`, which is 1 if male and 2 if female; and the feeling thermometer for then-President Barack Obama (`ftobama`), again from 0 to 100.

The `select()` function in the Tidyverse is very versatile. It can be combined with other functions like `starts_with()`, `ends_with()`, and `contains()` to select more than one variable at a time. We can also use the : to select more than one variable that are consecutive in the data set.

For example, if we wanted to select all of the feeling thermometer variables, and we know that they all start with the prefix `ft`, we could simply put the following.

```
# Select using starts_with()
NESdta %>%
  dplyr::select(starts_with("ft"))
```

Alternatively, since they are next to each other in the original data set, we could have done it like this.

```
# Selecting using a range
NESdta %>%
  dplyr::select(ftobama:ftsci)
```

We can also combine all of these tools. Let's say we wanted political party affiliation, year of birth, gender, and all of the feeling thermometer variables (with the prefix `ft*`).

```
# Combining selection procedures
NESdta %>%
  dplyr::select(pid3, birthyr, gender, starts_with("ft"))
```

```
## # A tibble: 1,200 x 21
```

	pid3	birthyr	gender	ftobama	ftblack	ftwhite	fthisp	ftgay
##	<dbl>	<dbl>	<dbl>	<dbl>	<dbl>	<dbl>	<dbl>	<dbl>
## 1	1	1960	1	100	100	100	100	96
## 2	3	1957	2	39	6	74	6	75
## 3	2	1963	1	1	50	50	50	16
## 4	1	1980	1	89	61	64	61	62
## 5	4	1974	1	1	61	58	71	55
## 6	3	1958	1	0	50	51	51	46
## 7	1	1978	1	73	100	70	100	100

```
## 8      2    1951      1      0      70      70      69   49
## 9      3    1973      1     12      50      50      50    5
## 10     1    1936      1     87      75      90      51   85
## # ... with 1,190 more rows, and 13 more variables:
## #   ftjeb <dbl>, fttrump <dbl>, ftcarson <dbl>,
## #   fthrc <dbl>, ftrubio <dbl>, ftcruz <dbl>,
## #   ftsanders <dbl>, ftfiorina <dbl>, ftpolice <dbl>,
## #   ftfem <dbl>, fttrans <dbl>, ftmuslim <dbl>, ftsci <dbl>
```

Finally, we can also use `select()` to remove columns by placing – in front of them. For example, if we decide we do not want to keep the 7 point political ID scale, we can remove it from the data set with the following code.

```
NESdta_short <- NESdta_short %>%
  dplyr::select(-pid7)


NESdta_short
```

```
## # A tibble: 1,200 x 6
##     fttrump pid3 birthyr ftobama state gender
##       <dbl> <dbl>  <dbl>   <dbl> <dbl>  <dbl>
## 1        1    1    1960     100     6      1
## 2       28    3    1957      39    13      2
## 3      100    2    1963       1    24      1
## 4        0    1    1980      89    35      1
## 5       13    4    1974       1    27      1
## 6       61    3    1958       0    18      1
## 7        5    1    1978      73    23      1
## 8       85    2    1951       0    53      1
## 9       70    3    1973      12    18      1
## 10       5    1    1936      87    12      1
## # ... with 1,190 more rows
```

The `filter()` function works similarly to the `select()` function, but instead of selecting columns by their *names*, `filter()` allows you to select rows by their *values*. If, for example, we wanted to find only the respondents who gave Donald Trump the highest possible rating, we could do this easily using this function. In this case, there were 54 people in the survey who matched this criterion. As with the `select()` function, we indicate that we want to use the `filter()` function from the `dplyr` package to avoid conflicts with other packages, `dplyr::filter()`.

```
# Select only those respondents who give Trump a 100
NESdta_short %>%
  dplyr::filter(fttrump == 100)


## # A tibble: 54 x 6
```

```
##     fttrump pid3 birthyr ftobama state gender
##       <dbl> <dbl>   <dbl>   <dbl> <dbl>  <dbl>
## 1       100     2    1963       1    24      1
## 2       100     4    1974       0    17      1
## 3       100     4    1947       3     6      2
## 4       100     3    1958       0    31      1
## 5       100     2    1936       6    48      1
## 6       100     2    1962       0    42      2
## 7       100     3    1957       1    45      1
## 8       100     2    1959       4    13      1
## 9       100     4    1952       0     4      2
## 10      100     2    1951       6    36      1
## # ... with 44 more rows
```

Let's say that we only want to see those respondents who give Donald Trump the highest possible rating (100) and Barack Obama the lowest possible rating (1). We can combine these conditions using the & ("and") operator.

```
# Filter respondents who give Trump a 100 and Obama a 1
NESdta_short %>%
  dplyr::filter(fttrump == 100 & ftobama == 1)
```

We could also look for those who either give Donald Trump the highest possible rating or give Barack Obama the highest possible rating by using the | ("or") operator. There are 144 respondents who match one of these two criterion.

```
# Filter respondents who give Trump a 100 or Obama a 100
NESdta_short %>%
  dplyr::filter(fttrump == 100 | ftobama == 100)
```

```
## # A tibble: 144 x 6
##     fttrump pid3 birthyr ftobama state gender
##       <dbl> <dbl>   <dbl>   <dbl> <dbl>  <dbl>
## 1        1     1    1960     100     6      1
## 2      100     2    1963       1    24      1
## 3       59     1    1945     100    42      2
## 4       16     1    1951     100    36      2
## 5       18     2    1994     100    40      2
## 6        1     1    1960     100    55      2
## 7      100     4    1974       0    17      1
## 8        0     1    1969     100    48      1
## 9        0     1    1936     100     4      1
## 10       6     3    1959     100    47      1
## # ... with 134 more rows
```

We can also combine these logical operators. Let's say that we want all the people who have either given both Donald Trump and Barack Obama scores of 100 or have given them both scores of 1. We can do this using parentheses, just like we would in a mathematical equation. This tells R to first find all the people who match the first criterion, then find all the people who match the second criterion, and select the people who match either one criterion or the other. Perhaps unsurprisingly, there are only four people who match one of these two criteria.

```
# Filter respondents for Trump and Obama 100s or 1s
NESdta_short %>%
  dplyr::filter((fttrump == 100 & ftobama == 100) |
                 (fttrump == 1 & ftobama == 1))
```

```
## # A tibble: 4 x 6
##    fttrump  pid3 birthyr ftobama state gender
##      <dbl> <dbl>   <dbl>   <dbl> <dbl>  <dbl>
## 1      100     1    1961     100    24      2
## 2        1     3    1981       1    26      1
## 3      100     1    1994     100    34      2
## 4        1     2    1981       1    45      2
```

Finally, we can also filter using ranges and other mathematical operators. If, for example, we wanted only those people whose approval of Donald Trump is greater than 50, we can do this much as you would expect.

```
# Filter respondents for Trump greater than 50
NESdta_short %>%
  dplyr::filter(fttrump > 50)
```

```
## # A tibble: 472 x 6
##     fttrump  pid3 birthyr ftobama state gender
##       <dbl> <dbl>   <dbl>   <dbl> <dbl>  <dbl>
## 1       100     2    1963       1    24      1
## 2        61     3    1958       0    18      1
## 3        85     2    1951       0    53      1
## 4        70     3    1973      12    18      1
## 5        74     2    1978      32    51      1
## 6        95     3    1943      10    36      2
## 7        82     2    1938      80    21      2
## 8        91     2    1956       4     6      2
## 9        51     3    1984       0     8      1
## 10       51     1    1981      66    39      1
## # ... with 462 more rows
```

This also applies to finding values that are related to the summary statistics for the full data. For example, if we wanted all of those who give a higher than average approval of Donald Trump, we could run the following.

```
NESdta_short %>%
  dplyr::filter(fttrump > mean(fttrump, na.rm = TRUE))
```

```
## # A tibble: 537 x 6
##    fttrump  pid3 birthyr ftobama state gender
##      <dbl> <dbl>   <dbl>   <dbl> <dbl>  <dbl>
## 1     100     2    1963       1    24      1
## 2      61     3    1958       0    18      1
## 3      85     2    1951       0    53      1
## 4      70     3    1973      12    18      1
## 5      74     2    1978      32    51      1
## 6      95     3    1943      10    36      2
## 7      82     2    1938      80    21      2
## 8      91     2    1956       4     6      2
## 9      51     3    1984       0     8      1
## 10     51     1    1981      66    39      1
## # ... with 527 more rows
```

Exercises

3.2.0.0.1 Easy

- Using the NESdta tibble, create a new tibble called NESdta_practice that only includes pid3 and fttrump.
- Using the NESdta tibble, create a new tibble that overwrites NESdta_practice that only includes variables containing the string pid.
- Filter the rows to only those respondents who give Donald Trump a higher approval than the median.
- How many people give Trump a score above 50 and Obama a score below 50?

3.2.0.0.2 Intermediate

- You can see that in some cases, we ran commands that actually change NESdta_short and in other cases we just printed what the data would look like. Why was the behavior different between these blocks of code? Explain how to change a data set in the computer's memory, how to create a new data set, and how to just print the results in the console.
- How would you select only variables in the NESdta data set that end with the number 3? Contains the string "id"?
- Type ?filter into the console. What do you get? Which help file would you choose, and why?

- Go through the data set and put together a tibble that selects five variables you would use to predict support for Trump. Check to make sure that the coding of these questions matches with what you would expect, and use `filter()` to remove missing values. [You will need to reference the data set codebook available online.]

3.3 Grouping and Summarizing Your Data

You may have noticed the strange symbol in a few of the previous commands, %>%. This is called a "pipe," which is read as "then," and it was originally developed by Stefan Milton Bache for the R package `magrittr`. The pipe allows you to declare at the very beginning the data on which you want to work and stack a number of operations onto that data without having to declare the data you want to use each time you issue a command (or, worse, only work with one data set at a time - yes, people really do this).

To show you how the pipe allows you to stack commands, let's look at two other functions - the `group_by()` and `summarize()` functions. Let's say that we think that Republicans will be most positively disposed to Donald Trump as a candidate, followed by Independents, and then by Democrats - not an earth-shaking hypothesis, but it works for demonstration. We can use the `group_by()` function to tell R what groups we want to make, and follow this with the `summarize()` command to create the needed summaries.[5]

```
# Using group_by() and summarize()
NESdta_short %>%
  group_by(pid3) %>%
  summarize(average_fttrump = mean(fttrump, na.rm = TRUE),
            median_fttrump = median(fttrump, na.rm = TRUE),
            variance_fttrump = var(fttrump, na.rm = TRUE),
            stddev_fttrump = sd(fttrump, na.rm = TRUE),
            max_fttrump = max(fttrump, na.rm = TRUE),
            min_fttrump = min(fttrump, na.rm = TRUE),
            n_fttrump = n()) %>%
  ungroup()
```

```
## # A tibble: 5 x 8
```

---

[5]The warning message about ungrouping output is a recent behavior change to the summarize() function. By default, the .group argument is set to "drop_last", meaning that the last, in this case only, grouping is dropped. We still use the ungroup() function at the end to avoid any unanticipated behavior.

```
##      pid3 average_fttrump median_fttrump variance_fttrump
##     <dbl>          <dbl>          <dbl>            <dbl>
## 1      1           22.1              4            4878.
## 2      2           65.1             72            1028.
## 3      3           44.5             41            3675.
## 4      4           45.7             41            1346.
## 5      5           28.8             26             874.
## # ... with 4 more variables: stddev_fttrump <dbl>,
## #    max_fttrump <dbl>, min_fttrump <dbl>, n_fttrump <int>
```

As you can see from the output, the group_by(pid3) function has declared that we want R to group respondents together by their party affiliation. The summarize(average_fttrump = mean(fttrump, na.rm = TRUE), ...) command is a little more complex. We are telling R that we are going to have it create a new variable, average_fttrump, that is the mean value of the variable for each group. The mean() function is a part of the base R system. The command na.rm = TRUE is necessary to tell R that we do not want it to include any missing values - "NA". Why is this needed? Well, technically the average of anything including a missing value is going to be missing. So we need to tell R explicitly that we do not want them to be included. The other functions - e.g. max(), min(), median() - work in a similar manner. Finally, since we are done with these groups, we run the ungroup() function. This ensures that the grouping does not persist past these commands.

Another useful function, but from base R, is the summary() function. This function takes an object as its input and outputs an adaptive display of summary statistics.

```
# Using the summary() function with a data set
summary(NESdta_short)
```

Similarly, we can use the summary() function on single variables using the $, where the data set is placed before the $ and the variable of interest is placed after, as shown below.

```
# Using the summary function with a variable
summary(NESdta_short$fttrump)
```

The group_by() command is also useful in a number of other conditions. In the last section, we filtered responses to only those who gave Donald Trump a higher than average approval score. This time, let's say that we are interested in finding the respondents of each gender who give him a higher than average score than other people of the same gender. In this case, we can tell R to group the responses by gender and then filter to those respondents who score higher than average in each group.

```
# Filter for Trump higher than average approval by gender
NESdta_short %>%
```

```
    group_by(gender) %>%
    dplyr::filter(fttrump > mean(fttrump, na.rm = T))
```

```
## # A tibble: 531 x 6
## # Groups:   gender [2]
##     fttrump  pid3 birthyr ftobama state gender
##       <dbl> <dbl>   <dbl>   <dbl> <dbl>  <dbl>
## 1      100     2    1963       1    24      1
## 2       61     3    1958       0    18      1
## 3       85     2    1951       0    53      1
## 4       70     3    1973      12    18      1
## 5       74     2    1978      32    51      1
## 6       95     3    1943      10    36      2
## 7       82     2    1938      80    21      2
## 8       91     2    1956       4     6      2
## 9       51     3    1984       0     8      1
## 10      51     1    1981      66    39      1
## # ... with 521 more rows
```

Finally, we can create the groups for any number of conditions. Extending our first example, let's say we want the mean approval of Donald Trump broken down by party affiliation and gender. This can be accomplished by just including both conditions, separated by a comma.

```
# Summarizing mean approval of Trump by party and gender
NESdta_short %>%
    group_by(pid3, gender) %>%
    summarize(average_fttrump = mean(fttrump, na.rm = TRUE),
              n_fttrump = n()) %>%
    ungroup()
```

```
## # A tibble: 9 x 4
##    pid3 gender average_fttrump n_fttrump
##   <dbl>  <dbl>           <dbl>     <int>
## 1     1      1            17.5       189
## 2     1      2            25.3       270
## 3     2      1            70.3       129
## 4     2      2            60.6       151
## 5     3      1            44.0       208
## 6     3      2            45.2       172
## 7     4      1            43.1        44
## 8     4      2            49.2        33
## 9     5      2            28.8         4
```

Exercises

3.3.0.0.1 Easy

- Using the `summary()` function, give the summary statistics for `ftobama`.
- Sometimes it is more useful to find out how many respondents fall within a category. Using the base R `table()` function find out how many people are in each category of the `pid3` variable.
- Using `group_by()` and `summarize()` find the summary statistics for `ftobama` by `gender`.
- When combining commands, we use the `%>%` (pipe). Try to put in your own words what it means to pipe data from one command to another. From the last example, what is being piped into each command? How is the data changed at each step?

3.3.0.0.2 Intermediate

- How would you do what you did in #18 using `group_by()` and `summarize()`?
- You have now seen the $ in a couple of situations. How would you describe (in words) the use of $?

3.3.0.0.3 Advanced

- Group the `NESdta_short` data object accordingly:
  - `group_by_all()` (all variables)
  - `group_by_at()` (using `pid3` and `gender`)
  - `group_by_if()` (for all *numeric* variables)

3.4 Creating New Variables

Another task that you will often find yourself doing is adding new variables to your data set. This is usually done with the `mutate()` function from the `dplyr` package in the Tidyverse. Let's start with a very simple variable transformation. The `birthyr` variable does not directly represent the concept we really want, age. To do this, we should create a new variable that calculates the age of a respondent by subtracting their birth year from the year of the survey.

```
# Create a new variable giving the respondent's age
NESdta_short <- NESdta_short %>%
  mutate(age = 2016 - birthyr)


NESdta_short

## # A tibble: 1,200 x 7
```

```
##      fttrump  pid3 birthyr ftobama state gender  age
##       <dbl> <dbl>   <dbl>   <dbl> <dbl>  <dbl> <dbl>
## 1         1     1    1960     100     6      1    56
## 2        28     3    1957      39    13      2    59
## 3       100     2    1963       1    24      1    53
## 4         0     1    1980      89    35      1    36
## 5        13     4    1974       1    27      1    42
## 6        61     3    1958       0    18      1    58
## 7         5     1    1978      73    23      1    38
## 8        85     2    1951       0    53      1    65
## 9        70     3    1973      12    18      1    43
## 10        5     1    1936      87    12      1    80
## # ... with 1,190 more rows
```

This works for any number of different operations on variables. For example, if we wanted to get the square of the respondent's age, we could simply do the following.

```
# Create a new variable for squared respondent's age
NESdta_short <- NESdta_short %>%
  mutate(age2 = age^2)
```

```
NESdta_short
```

```
## # A tibble: 1,200 x 8
##      fttrump  pid3 birthyr ftobama state gender  age  age2
##       <dbl> <dbl>   <dbl>   <dbl> <dbl>  <dbl> <dbl> <dbl>
## 1         1     1    1960     100     6      1    56  3136
## 2        28     3    1957      39    13      2    59  3481
## 3       100     2    1963       1    24      1    53  2809
## 4         0     1    1980      89    35      1    36  1296
## 5        13     4    1974       1    27      1    42  1764
## 6        61     3    1958       0    18      1    58  3364
## 7         5     1    1978      73    23      1    38  1444
## 8        85     2    1951       0    53      1    65  4225
## 9        70     3    1973      12    18      1    43  1849
## 10        5     1    1936      87    12      1    80  6400
## # ... with 1,190 more rows
```

And, if we wanted to get rid of that same variable later, we could do that by setting its value to NULL.

```
# Remove variable with square of age from data set
NESdta_short <- NESdta_short %>%
  mutate(age2 = NULL)
```

```
NESdta_short
```

```
## # A tibble: 1,200 x 7
##     fttrump  pid3 birthyr ftobama state gender  age
##       <dbl> <dbl>   <dbl>   <dbl> <dbl>  <dbl> <dbl>
##  1        1     1    1960     100     6      1    56
##  2       28     3    1957      39    13      2    59
##  3      100     2    1963       1    24      1    53
##  4        0     1    1980      89    35      1    36
##  5       13     4    1974       1    27      1    42
##  6       61     3    1958       0    18      1    58
##  7        5     1    1978      73    23      1    38
##  8       85     2    1951       0    53      1    65
##  9       70     3    1973      12    18      1    43
## 10        5     1    1936      87    12      1    80
## # ... with 1,190 more rows
```

In the last section, we summarized support for then-candidate Donald Trump by party affiliation. But what if we want these summaries to be a part of the NESdta_short data set? This is where the mutate() function comes in. Run the same functions as above, but this time let us use the mutate() function instead of the summarize() function.

```
# Creating a new variable using group_by() and mutate()
NESdta_short <- NESdta_short %>%
  group_by(pid3) %>%
  mutate(average_fttrump = mean(fttrump, na.rm = TRUE)) %>%
  ungroup()

NESdta_short
```

As you can see, a sixth column has been added to our data set, with the average values for each political ID added to the data set. From here, we can take other actions. For example, we can subtract the average for each group from the individual respondent's evaluation of candidate Donald Trump by using the mutate() function again.

```
# Using mutate to create a new variable
NESdta_short <- NESdta_short %>%
  mutate(deviation_fttrump = fttrump - average_fttrump)
NESdta_short
```

A new column has been added showing how far away each respondent is from the average for those who share their party affiliation. Respondent 1, shown in the first row, gives Donald Trump a rating about 21 points lower than the average for those who share their party affiliation.

Note that while the feeling thermometers for Donald Trump and Barack

Obama are only supposed to go from 0 to 100, the summary statistics said
the maximum values were 998. What is happening here?

Many data sets try not to leave blank spaces or mix strings and numeric values.
The reason is that some programs might behave unexpectedly when loading
this data. So, instead, they represent missing values by highly improbable
numeric values – in this case 998 (other data sets will use unexpected negative
values like -999). We need to tell R that these are actually missing values,
denoted as NA in R, as opposed to actual numbers.

To do this, we can again use the `mutate()` function. This time, we combine it
with the `replace()` function. `replace()` takes three values as its input. The
first is the variable on which we are making the replacement, the second is
a logical test. This can be read as, "Where the variable is..." For example,
the second part of the first replacement asks it to make the replacement
where the variable `fttrump` is greater than 100. As you can see, within the
`mutate()` function, we have asked for our original variable to be equal to the
specified replacement (i.e., we have *redefined* the original variable to drop these
nonsensical values).

```
# Using replace() to recode values
NESdta_short <- NESdta_short %>%
  mutate(fttrump = replace(fttrump, fttrump > 100, NA),
         ftobama = replace(ftobama, ftobama == 998, NA))
summary(NESdta_short)
```

```
##     fttrump          pid3           birthyr
## Min.   :  0.00   Min.   :1.000   Min.   :1921
## 1st Qu.:  2.00   1st Qu.:1.000   1st Qu.:1955
## Median : 30.00   Median :2.000   Median :1968
## Mean   : 38.38   Mean   :2.072   Mean   :1968
## 3rd Qu.: 72.00   3rd Qu.:3.000   3rd Qu.:1982
## Max.   :100.00   Max.   :5.000   Max.   :1997
## NA's   :3
##     ftobama          state           gender
## Min.   :  0.00   Min.   : 1.00   Min.   :1.000
## 1st Qu.:  5.00   1st Qu.:12.00   1st Qu.:1.000
## Median : 52.50   Median :29.00   Median :2.000
## Mean   : 48.62   Mean   :28.32   Mean   :1.525
## 3rd Qu.: 87.00   3rd Qu.:42.00   3rd Qu.:2.000
## Max.   :100.00   Max.   :56.00   Max.   :2.000
## NA's   :2
##       age
## Min.   :19.00
## 1st Qu.:34.00
## Median :48.00
```

```
## Mean   :48.06
## 3rd Qu.:61.25
## Max.   :95.00
##
```

Another variable we will likely want to change is the `state` variable. Right now, it has numbers that represent the states, but we will probably want strings with the state names as well. We can look up the numbers associated with each state in the ANES and create a new variable called `state_name` that contains the name of the state.

There are a lot of values we will need to replace, so we will use a different function, the `case_when()` function, which allows us to change a large number of values within a variable.

```
# Create state_name with the string names of states
NESdta_short <- NESdta_short %>%
  mutate(state_name = case_when(state == 1~"Alabama",
                                state == 2~"Alaska",
                                state == 4~"Arizona",
                                state == 5~"Arkansas",
                                state == 6~"California",
                                state == 8~"Colorado",
                                state == 9~"Connecticut",
                                state == 10~"Delaware",
                                state == 11~"DC",
                                state == 12~"Florida",
                                state == 13~"Georgia",
                                state == 15~"Hawaii",
                                state == 16~"Idaho",
                                state == 17~"Illinois",
                                state == 18~"Indiana",
                                state == 19~"Iowa",
                                state == 20~"Kansas",
                                state == 21~"Kentucky",
                                state == 22~"Louisiana",
                                state == 23~"Maine",
                                state == 24~"Maryland",
                                state == 25~"Massachusetts",
                                state == 26~"Michigan",
                                state == 27~"Minnesota",
                                state == 28~"Mississippi",
                                state == 29~"Missouri",
                                state == 30~"Montana",
                                state == 31~"Nebraska",
                                state == 32~"Nevada",
```

```
                            state == 33~"New Hampshire",
                            state == 34~"New Jersey",
                            state == 35~"New Mexico",
                            state == 36~"New York",
                            state == 37~"North Carolina",
                            state == 38~"North Dakota",
                            state == 39~"Ohio",
                            state == 40~"Oklahoma",
                            state == 41~"Oregon",
                            state == 42~"Pennsylvania",
                            state == 44~"Rhode Island",
                            state == 45~"South Carolina",
                            state == 46~"South Dakota",
                            state == 47~"Tennessee",
                            state == 48~"Texas",
                            state == 49~"Utah",
                            state == 50~"Vermont",
                            state == 51~"Virginia",
                            state == 53~"Washington",
                            state == 54~"West Virginia",
                            state == 55~"Wisconsin",
                            state == 56~"Wyoming"))
```

A final note: you might have noticed the double equal sign, ==. This is a relatively common logical operator used in many statistical packages and programming languages. A single equal sign, =, is used to set one object equal to another. So, in the command above, when we type `fttrump = ...`, this tells R to change the object `fttrump` into what follows the equal sign. A double equal sign, ==, is used for comparison, and it returns a value of TRUE if the item on the left-hand side is equal to the item on the right-hand side, and FALSE otherwise.

You will use this a lot, especially as we start discussing the use of logic. A type of logical command you will find yourself using a lot is `ifelse(condition, outcome if true, outcome if false)`. Let's take, for example, the gender variable in the ANES data. Here, we are interested in recoding the gender variable (currently 2 = female and 1 = male) to be more descriptive and also on the more common 0,1 scale. Using `mutate()` and `ifelse()` from base R, we create a new variable female, where 1 equals cases when gender = 2 (female), and 0 otherwise (previously, gender = 1).

```
# Gender is currently coded 1 for male 2 for female
unique(NESdta_short$gender)
```

```
## [1] 1 2
```

```
# Use ifelse() to create a dichotomous variable if female
NESdta_short <- NESdta_short %>%
  mutate(female = ifelse(gender == 2, 1, 0))
```

`ifelse()` is a very flexible function. It can be used to execute multiple logical statements by nesting those statements (an approach we will see again later in the *Essential Programming* chapter). To nest these functions, we simply tell the computer that if the outcome is false, it is to execute another `ifelse()` function. Let's say we wish to split the `age` variable into three categories young, `middle aged`, and `old`. We can do this using nested `ifelse()` functions.

```
#Using nested ifelse() functions
NESdta_short %>%
  mutate(age_categories = ifelse(age <= 35, "Young",
                          ifelse(age > 35 & age < 65,
                                 "Middle Age", "Old"))) %>%
  group_by(age_categories) %>%
  summarize(n = n())
```

```
## # A tibble: 3 x 2
##   age_categories     n
##   <chr>          <int>
## 1 Middle Age       649
## 2 Old              210
## 3 Young            341
```

In the above code block, we nest two `ifelse()` functions. The first tests if the respondent's age is less than or equal to 35. If true, it assigns a value of "Young"; if false it goes to the next test. The second `ifelse()` function asks if the respondent's age is between 35 and 65. If it is, the respondent is assigned a value of "Middle Age", and, if not, they must be "Old". The last two lines utilize the `group_by()` and `summarize()` functions you learned about above to show how many people in our survey fall into each category.

Exercises

3.4.0.0.1   Easy

- Create a new variable called Republican that is 1 if the respondent is a Republican (`pid3 == 2`) and 0 otherwise.
- Create a new variable called `pid_text` that gives the text labels for `pid3` (1 = Democrat, 2 = Republican, 3 = Independent, 4 = Other).

3.4.0.0.2   Intermediate

- Use `replace()` to change those who are labeled "Independent" in your `pid_text` variable to "Other."

- Create a new variable that is the de-meaned version of `ftobama`. Try to do it in one step using `%>%`.

3.4.0.0.3 Advanced

- Mutate a new variable of your choice, but it must be the combination of *three* other variables. Consider using `case_when()`, among other useful `dplyr` functions.
- Create a `tibble` and a `tribble` of the most recent `NESdta_short` data object. What are the differences, and what do these differences substantively point to in the Tidyverse? How might they compare to a `data.frame`?

3.5 Combining Data Sets

One area where R really shines is in its ability to handle multiple data sets at the same time. In many other common statistical programs, you are limited to working on one data set at a time within a particular session. In R, you can work with many more. Actually, right now we are already working with more than one data set. When we created the `NESdta_short` data set, we added a second data set to our session. If you look in the upper-right hand window of RStudio, under the Environment tab, you will see that both `NESdta` and `NESdta_short` are listed as "Data." This means that at any time you can go back to working on the original `NESdta` data set at any time.

Suppose we forgot a variable we wanted when we created the `NESdta_short` data set. All we would need to do is go back to the line where we did the subsetting above and run it with the additional variable name. No harm, no foul. For those of you who have worked with other statistics programs, you have probably seen what a pain similar operations can be.

But the primary use you will have for this is to work with different data sets. Say we suspect that where a person lives affects their approval of then-candidate Donald Trump - there were certainly differences in his voteshare in different states during the primary, and we know that people's political opinions are not independent of those around them. We can load a data set with a few state-level attributes and combine it with our individual-level ANES data.

```
# Read states data set into a tribble
states <- read_csv(here("Data","statescsv.csv"))
```

To merge two data sets, we need to find a common key. This key is a variable that links cases in one data set to another. In this case, we have the `state_name` variable we created in the ANES data set and the variable state in our states data set. Both contain strings with the name of the state in them.

It will be useful to have the variables we are using as the common key to have the same name. If we ever need to do this, we can use the `rename()` function, with the new name on the left and the old name on the right. For this data, we need to rename the state variable to state_name to match with the ANES data we just created.[6]

```
# Change "state" to "state_name" to match ANES data set
states <- states %>%
  rename(state_name = state)
```

Now we will join the data sets together. There are several options for joining data together. These differ in how they handle situations in which the data sets have somewhat different values in the common key. Let's say, for example, that data set #1 includes Puerto Rico and data set #2 does not. Conversely, data set #2 includes Guam and data set #1 does not.

We can decide:

1. To only *keep* those cases where the territory is *common to both data sets*, thereby excluding both Puerto Rico and Guam. This is an `inner_join()`.
2. To *keep* all the values in data set *#1* and *drop* the values that do not match in data set *#2* - in this case to keep Puerto Rico and drop Guam. This is a `left_join()`.
3. To *keep* all the values in data set *#2* and *drop* the values that do not match in data set *#1* - in this case to keep Guam and drop Puerto Rico. This is a `right_join()`.
4. To *keep* all the values from both *data set #1 and data set #2* - keeping both Puerto Rico and Guam. This is a `full_join()`.
5. To *drop* all values that match in *data set #1 and data set #2* - in this case, *only* keeping Puerto Rico and Guam. This is an `anti_join()`.

For this case, we will use an `inner_join()`, only keeping the values for the states that match in both data sets, and putting them into a new data set called `NESdta_states`.

```
# Create new data set by inner joining the NES and states data
NESdta_states <- NESdta_short %>%
  inner_join(states, by = "state_name")
# Display the variable names of the resulting data set
names(NESdta_states)

## [1] "fttrump"    "pid3"        "birthyr"      "ftobama"
```

---

[6]We can also do this without renaming the key variable.

```
## [5] "state"      "gender"    "age"      "state_name"
## [9] "female"     "college"   "over64"   "south"
## [13] "unemploy"   "union"     "urban"
```

As we can see, the resulting data set has all the variables from both data sets.

Exercises

3.5.0.0.1 Easy

- Another way to combine two data sets with different names for their key, instead of renaming one of them, is to use by = c("key_name_1" = "key_name_2"). Reload the state data set and try this out. What happens with the key names?
- What is anti_join(), and when might you use this?

3.5.0.0.2 Intermediate

- Reload the states data set. What happens when you use outer_join() instead of inner_join()? Why? How might this behavior change in different circumstances?

3.5.0.0.3 Advanced

- Sometimes two data sets will have the same name for a non-key variable. What do you think happens in this case? Reload the states data set, change the name of demstate in the data to pid3. Merge the data sets and use the name() function to see what happened, and report your results with some brief discussion.

3.6 Basic Descriptive Analysis

Now that we have seen how to load our data and do some basic manipulation, you might be interested in describing your data in a few different ways that facilitate testing of hypotheses. In this section, we will cover some common descriptive methods for characterizing relationships in your data. This will also give us an opportunity to play around with some of the data manipulation commands you learned above. We will start with a discussion of cross-tabulation and then move into some variations of comparisons of means.

Before beginning this analysis, let's make a few changes to the data using the tools we learned above to make our results more interpretable. Take some time to look at this block to make sure you understand what we have covered so far.

```
NESdta_states <- NESdta_states %>%
  mutate(gender_name = ifelse(gender == 1, "male", "female"),
         pid_name = case_when(pid3 == 1 ~ "Democrat",
                              pid3 == 2 ~ "Republican",
                              pid3 == 3 ~ "Other"),
         south_name = ifelse(south == 1, "South", "Not South"),
         age_categories = case_when(age <= 35 ~ "Young",
                                    age > 35 & age <= 65 ~
                                      "Middle Age",
                                    age > 65 ~ "Older"))
```

Cross-tabulations simply compare how many cases fall into different groups with two categorical or ordered variables. In our current data set, we have two categorical variables about which we might want information, gender and political ID. Suppose we expected women to be more likely than men to support Democrats. Let's see if this holds true in the NES data set.

There are several ways we might try to accomplish this. One is to use the same summary tools that we used above. We can simply use the group_by() function to find out how many fall into each category of these variables.

```
NESdta_states %>%
  group_by(gender_name, pid_name) %>%
  summarize(n = n())
```

```
## # A tibble: 8 x 3
## # Groups:   gender_name [2]
##   gender_name pid_name        n
##   <chr>       <chr>       <int>
## 1 female      Democrat      270
## 2 female      Other         172
## 3 female      Republican    151
## 4 female      <NA>           37
## 5 male        Democrat      189
## 6 male        Other         208
## 7 male        Republican    129
## 8 male        <NA>           44
```

This process produces all of the information that we would expect in a cross-tabulation (although not exactly in the format we might expect). For example, we can see that there are 129 men who are Republicans in this survey. But this does not necessarily answer our main question, since there are also 151 women who are Republicans. If we go by raw numbers, we might assume women are more likely to be Republicans, missing the fact that there are also more women in the survey overall.

To get this, we will need to figure out how many men and women are in this survey, and then divide the numbers above by the total number of men and women. This is a straightforward combination of the commands we have already used above – first grouping by gender and getting the number in each category, then grouping by both gender and political ID and getting those numbers, and finally dividing the former by the latter. The only last piece is that, since we do not want to lose our other variables, we will use mutate for the first grouping. We will also use mean() to get total number by each gender (technically, we could have used min(), max() or any other summary function because the values are all the same within this grouping). Finally, we divide the number of observations in each gender/political ID pair by the total number of respondents of each gender to get the proportion.

```
NESdta_states %>%
  group_by(gender_name) %>%
  mutate(sum_gender = n()) %>%
  group_by(gender_name, pid_name) %>%
  summarize(n = n(),
            n_gender = mean(sum_gender),
            p = n/n_gender)
```

```
## # A tibble: 8 x 5
## # Groups:   gender_name [2]
##   gender_name pid_name        n n_gender       p
##   <chr>       <chr>       <int>    <dbl>   <dbl>
## 1 female      Democrat      270      630  0.429
## 2 female      Other         172      630  0.273
## 3 female      Republican    151      630  0.240
## 4 female      <NA>           37      630  0.0587
## 5 male        Democrat      189      570  0.332
## 6 male        Other         208      570  0.365
## 7 male        Republican    129      570  0.226
## 8 male        <NA>           44      570  0.0772
```

The results confirm that there is a gender difference, but, at least in this sample, the main difference appears to be in terms of the proportion of each gender that identifies as Democrat versus as a member of neither party. About 43% of women identify as Democrats, while only 33% of men do the same. Conversely, 36% of men say they do not identify with either party, whereas 27% of women say the same. The proportion of Republicans in both groups is pretty similar, with only a 1.3 point difference. We can repeat this process with three or more variables if desired.

While the above provides all the information you need for a cross-tabulation, we will admit that it is not the prettiest way to do things. Perhaps not surprisingly, then, a package has been built to make this process even easier. The janitor package gives us the ability to create these types of cross-tabulations very easily. We use three simple functions: `tabyl()` is where we put the variables we want cross-tabulated, `adorn_percentages()` allows us to choose if we want "row" or "column" proportions, `adorn_pct_formatting()` converts the proportions into percents and allows us to set the number of digits, and `adorn_ns()` results in the inclusion of the raw counts.

In the block below, we load the `janitor` package and create a simple cross-tabulation of the number of outcomes in each category.

```
library(janitor)

NESdta_states %>%
  tabyl(pid_name, gender_name)

##     pid_name female male
##     Democrat    270  189
##        Other    172  208
##   Republican    151  129
##         <NA>     37   44
```

In the next block, we add in the `adorn_percentages()` function and tell it to give us "col" (column) proportions.

```
NESdta_states %>%
  tabyl(pid_name, gender_name) %>%
  adorn_percentages("col")

##     pid_name      female        male
##     Democrat  0.42857143  0.33157895
##        Other  0.27301587  0.36491228
##   Republican  0.23968254  0.22631579
##         <NA>  0.05873016  0.07719298
```

Finally, we convert the proportions into percentages and include the counts to create a nicely formatted cross-tabulation.

```
NESdta_states %>%
  tabyl(pid_name, gender_name) %>%
  adorn_percentages("col") %>%
  adorn_pct_formatting(digits = 2) %>%
  adorn_ns()

##     pid_name        female          male
##     Democrat  42.86% (270)  33.16% (189)
##        Other  27.30% (172)  36.49% (208)
##   Republican  23.97% (151)  22.63% (129)
##         <NA>   5.87%  (37)   7.72%  (44)
```

The functions in the `janitor` package for creating cross-tabulations are quite flexible. If we want to create a three-way table, we can do this by simply adding the third variable we wish to include. For example, we can generate a cross-tabulation splitting the relationship between gender and partisan ID by whether respondents live in a southern or non-southern state.

```
NESdta_states %>%
  tabyl(pid_name, gender_name, south_name) %>%
  adorn_percentages("col") %>%
  adorn_pct_formatting(digits = 2) %>%
  adorn_ns()
```

```
## $`Not South`
##    pid_name       female         male
##    Democrat 45.70% (186) 34.85% (130)
##       Other 26.29% (107) 37.00% (138)
##  Republican 20.88%  (85) 18.77%  (70)
##        <NA>  7.13%  (29)  9.38%  (35)
##
## $South
##    pid_name      female        male
##    Democrat 37.67% (84) 29.95% (59)
##       Other 29.15% (65) 35.53% (70)
##  Republican 29.60% (66) 29.95% (59)
##        <NA>  3.59%  (8)  4.57%  (9)
```

Now, let's look at how to create tables to compare the means of a continuous variable within a category. This is very easy to do using the data manipulation functions we learned above. We can use `group_by()` to set our categories and `summarize()` to calculate the means of our target variable. Let's, for example, look at the differences between men and women in their approval of Donald Trump.

```
NESdta_states %>%
  group_by(gender_name) %>%
  summarize(averge_Trump_approval = mean(fttrump, na.rm = T))
```

```
## # A tibble: 2 x 2
##   gender_name averge_Trump_approval
##   <chr>                       <dbl>
## 1 female                       35.9
## 2 male                         41.1
```

Surprisingly, we do not see much difference in average ratings. Still, it should be noted that this poll took place prior to numerous allegations of sexual harassment and assault being lodged against then-candidate Donald Trump.

Again, this can easily be extended to more than one category by adding more than one group. Let's demonstrate this using by comparing women and men in different age categories.

```
NESdta_states %>%
  group_by(age_categories, gender_name) %>%
  summarize(averge_Trump_approval = mean(fttrump, na.rm = T))
```

```
## # A tibble: 6 x 3
## # Groups:   age_categories [3]
##   age_categories gender_name averge_Trump_approval
##   <chr>          <chr>                       <dbl>
## 1 Middle Age     female                       36.3
## 2 Middle Age     male                         43.2
## 3 Older          female                       46.3
## 4 Older          male                         52.4
## 5 Young          female                       28.7
## 6 Young          male                         31.2
```

This shows a very interesting pattern at this point in the election cycle. There are relatively large gender differences, but they seem to be dependent on age categories. The gap between men and women is quite apparent, with young women being more approving than their male counterparts, while middle-aged and older women are less approving than their male counterparts. As noted above, this likely changed later in the election cycle as more information came to light, but, in this relatively early period, it seems to be the women who remember Trump's tabloid history with women in the 1980s who have a lower approval than males in their age cohort.

3.7   Tidying a Data Set

So far we have primarily looked at the Tidyverse functions associated with the dplyr package. Another important data munging package in the Tidyverse is tidyr. The tidyr package is meant to assist in creating a "tidy" data set. Formulated by Hadley Wickham (Wickham, 2014), there are three rules that make a data set tidy: (1) each variable must have its own column, (2) each observation must have its own row, and (3) each value must have its own cell.

A commonly used, and very *untidy*, dataset is the World Development Indicators from the World Bank. We will load a raw output from the World Development Indicators in the same form you would receive if you used their interactive website.

```
# Read in World Development Indicators data set
wdi_data <- read_csv(here("data","wdi_data.csv"))
wdi_data
```

This very short data is very untidy for a number of reasons. First, our main variables, access to electricity and agricultural land, are not given their own columns, but are rather separate rows for each country. Second, each observation does not have its own row. Instead, we have the yearly observations in separate columns. Indeed, the only rule of tidy data sets that this data set follows is that each value has its own cell.

To begin the process of tidying this data, we make some changes to get it ready to reshape. First, we give it more usable variable names using the `rename()` function. You will notice that when we have variable names that are more than one word or start with numeric values, we have to surround them with "'", this is to indicate that these are not numeric or separate values. After we have renamed the columns we want, we also create a new variable with the variable names in the rows using the `mutate()` function. Finally, we get rid of the original row variable labels by deselecting them.

```
# Prepare WDI data for reshaping
wdi_data2 <- wdi_data %>%
  rename(country = `Country Name`, code = `Country Code`,
         series = `Series Name`, series_code = `Series Code`,
         `2010` = `2010 [YR2010]`, `2013` = `2013 [YR2013]`) %>%
  mutate(variable_name=case_when(series_code=="EG.ELC.ACCS.ZS"~
                                  "electricity_access",
                                 series_code=="AG.LND.AGRI.ZS"~
                                  "pct_agriculture")) %>%
  dplyr::select(-series, -series_code)
```

To get the years into their own column, we will take those two columns and use the `gather()` function on them. The `key` will be the name given to the new variable containing the column names and the `value` will be the name for the values in those columns. A common way to describe this process is that we have taken a "wide" data set and made it "long".

```
# Reshape the data wide to long
wdi_data2 <- wdi_data2 %>%
  gather(`2010`,`2013`, key = "year", value = "levels")
wdi_data2
```

We could also have indicated a range of columns to be gathered by using a colon (:) as an operator indicating, "from here to there." This is useful if we have a large number of columns to be gathered (as long as they are sequential in the data set).

```
# Reshape the data wide to long
wdi_data2 <- wdi_data2 %>%
  gather(`2010`:`2013`, key = "year", value = "levels")
wdi_data2
```

Now we want to give each variable its own column. To do this, we will use the `spread()` function. In this case, the `key` is he column we want to spread and the `value` is the variable level for those keys. This is the opposite of what we did with `gather()`. We are now taking a "long" data set and making it "wide".

```
# Reshape data long to wide
wdi_data2 <- wdi_data2 %>%
  spread(key = variable_name, value = levels)
wdi_data2
```

Now we have a tidy data set with which to work.

Exercises

3.7.0.0.1   Easy

- Explain/think about the differences between `spread()` and `gather()`. What are some common features? Unique features? When should one be used over the other and why?

3.7.0.0.2   Intermediate

- The `spread()` function can also be used to organize the summary analyses (cross-tabulations and comparison of means) that we created above using `group_by()`. Take the cross-tabulation created using `group_by()` into a format closer to what we created using the `janitor` package's `tabyl()` function using `spread()`.
- Do the same as in #1, but with the comparison of means.

3.7.0.0.3   Advanced

- Can you undo what we just did with the WDI data?

3.8   Saving Your Data Set for Later Use

After all the work you have done to get your data into the shape you want, you will probably want to save this data set to your hard drive so you do not have to start over in your next session. To do this, we recommend using the `write_csv()` function from the `readr` package in Tidyverse.

There are several reasons we recommend saving your data. First, we suggest saving data as a `.csv` file because text-based storage files like this are quite compact, can be opened by a range of programs and languages, and will not become obsolete in the future. Older users of Stata or SPSS can attest that using proprietary storage can results in loss of data once the program manager decides to update the software and not maintain backward compatibility. Second, much like the difference between `read_csv()` from `readr` and base R's `read.csv()` function, the Tidyverse version has some defaults that users are likely to prefer. For example, the base R command (`write.csv()`) adds row names to the data set by default with no variable name. We have yet to encounter a situation in which this adds value to the data set and can sometimes cause problems, especially on data sets that are repeatedly opened and modified.

Saving your data set is relatively simple. You simply add two arguments to the `write_csv()` function. The first is the tibble or `data.frame` you wish to save. The second is where you want it saved, including the file name you wish to use. Here we are going to save our `NESdta_short` tibble as a `.csv` file called `ANES_short.csv` in our data folder.

```
# Save the NESdta_short tibble as a .csv file
write_csv(NESdta_short, here("data", "ANES_short.csv"))
```

As you might expect, there are `write_*` versions of all of the `read_*` commands used earlier for loading data. This makes R very flexible for opening and converting a wide variety of data sources.

Exercises

3.8.0.0.1 Easy

- Can you save this in `.dta` (Stata) format? Which package would you use?
- Can you save this in `.sav` (SPSS) format? Which package would you use?
- Can you save this in `.xlsx` (Excel) format? Which package would you use?

3.9 Saving Your Data Set Details for Presentation

Once you have all the data you need for your analysis in the format that you want, it is time to save that information in a format that you can use to present it in a paper or book. We have all been in the situation where we have put in a ton of work putting together a data set and a reviewer catches a small error or suggests the addition of a new variable. At one point in time, we would have manually typed in all the numbers and formatting, requiring that even some relatively minor changes resulted in hours of extra work. As you might already suspect, when there are problems like these, R users have likely written a package for dealing with the issue.

The `stargazer` package was designed so that you can easily take your analyses and turn them into professional tables that can be inserted into a word processing document (e.g. Word, LibreOffice, LaTeX). It will take care of formatting, updating, and most of the other tasks with little work on your part. Not only that, but `stargazer` is extremely flexible – able to accommodate a wide range of table formats, custom standard errors, and other quirks you may encounter in particular journals or with particular reviewers.

Here we will show how `stargazer` produces a table of summary statistics that can be inserted into a Microsoft Word document. Later, we will show how to generate a table for regression models.

We will start by making sure we have an object that only includes the columns we wish to summarize. In this case, let's just pick 3 variables: `fttrump`, `age`, and whether the respondent is `female`. We create the `age` variable by subtracting the year of the survey the respondent's year of birth (`birthyr`) from the year of the survey, 2016. We will create a dummy variable indicating whether the respondent is female using the same `ifelse()` statement we used above. The we will use `select()` to pick just those three columns.

```
# Create and select variable to be summarized
NESdta_summary <- NESdta_short %>%
  mutate(age = 2016 - birthyr,
         female = ifelse(gender == 2, 1, 0),
         fttrump = replace(fttrump, fttrump > 100, NA)) %>%
  dplyr::select(fttrump, age, female)
```

Now that we have the data set to be summarized, we can load the `stargazer` library and run the `stargazer()` function on the data set. Note that we need to convert our tibble to a `data.frame` for stargazer.

```
# Load stargazer package into workspace
library(stargazer)
```

```
# Create LaTeX-style table to print to console
stargazer(data.frame(NESdta_summary))
```

For those of you not familiar with LaTeX, the output might look a little strange. LaTeX is a document preparation system to produce high-quality typesetting. It is commonly used by academics because of its ability to automate some parts of the writing process (e.g. creating a formatted bibliography). It can also be used to automatically update tables and figures from R. It also, however, has a somewhat steep learning curve, so we will not assume you use it here.

Instead, let's create an HTML table. These can be opened natively in Microsoft Word and simply copied and pasted into any document. To do this, we will set the type of chart to HTML.

```
# Create stargazer table in .html format
stargazer(data.frame(NESdta_summary),
          type = "html")
```

This still looks confusing, but Microsoft Word (and most other visual word processing programs) knows how to read this to form a table. All we need to do is save it and open it using Word and it will look like a well-formatted table. To do this, we simply specify where to put the output and save it as a .doc file. In this case, we have created a sub-directory in our working directory for tables, and we will call our file "summary_table.doc".

```
# Save the table as a summary_table.doc
stargazer(data.frame(NESdta_summary),
          type = "html",
          out = here("tables","summary_table.doc"))
```

Now we have a well-formatted, easy to modify and read table. But there is one last thing we might want to change. The variable names in our data set are not very informative. We might want to make them a little clearer in meaning. We can do this by adding a vector of covariate labels, which is a collection of names bound together by the function c().

```
# Add informative variable labels
stargazer(data.frame(NESdta_summary),
          type = "html",
          covariate.labels = c("Approval of Trump",
                               "Age",
                               "Female"),
          out = here("tables","summary_table.doc"))
```

If you open "summary_table.doc" in Microsoft Word, you will see an output like that in the Figure 3.1. This output can be modified using Word's standard table manipulation tools and can be copied and pasted into any other Word document. For users of LaTeX, the process is even simpler. The user can save the table as a .tex file and add \input{./Tables/summary_table.tex} to their document. This will also allow for tables to be automatically updated as updates are made to your R code.

stargazer is very flexible and rich, with many options for customizing your tables. And once you have written the code for your table once, all you need to do in order to update it is make a small modification and re-run the code.

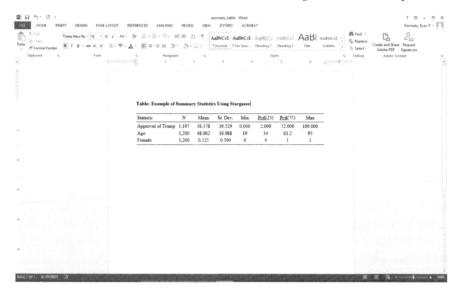

FIGURE 3.1
An Example of a Microsoft Word Summary Statistics Table Created by Stargazer

To learn more about the types of table `stargazer` can make and how to vary features, check out the package's online documentation.

Exercises

3.9.0.0.1 Easy

- List the arguments in the `stargazer()` function, and highlight a few that seem particularly useful and briefly describe how you might use this in your own research.
- Change the variable names in the `stargazer()` function and re-write a new table. Save the table on your Desktop as a .html file.

3.9.0.0.2 Intermediate

- Select all the feeling thermometer questions (they start with `ft`) from the ANES survey. Create a table of summary statistics from this for Microsoft Word. What do these tell you?

3.9.0.0.3 Advanced

- Replicate the stargazer output *without* calling the function.

4

Visualizing Your Data

Creating effective visual representations of data is arguably one of the most important parts of presenting research. Major social science journals, such as the *American Journal of Political Science*, have issued recent statements strongly recommending authors offer visual output in lieu of redundant numerical output whenever possible. Further, the importance of findings and novelty of design can be lost (or at best *limited*) if the researcher fails to offer simple, clear visualization of findings. As a group of scholars argued in the *Proceedings of the National Academy of Sciences*, the ability to read and construct data visualizations is "as important as the ability to read and write text" (Börner et al., 2019).

Given this importance, this chapter focuses on moving from creating "less-than-exciting" plots to exciting plots using the powerful `ggplot2` package, which is a core part of the Tidyverse. We will be covering all the basic plots and showing how, by learning the "grammar of visualization" associated with `ggplot2`, you can build quite complex and informative plots. We will also give you a taste of how to produce plots that are interactive and can be placed, for example, on a website to promote your research.

Throughout this chapter, we will also be providing examples of how plots are created in base R. The idea is to give you an idea of how plots are usually generated, and why associating a *grammar* with your plots is useful.

4.1 The Global Data Set

While the previous chapter focused on understanding political behavior in the 2016 American presidential primary, this chapter shifts its focus to understanding the relationship between economic development and democracy from 1972 to 2014. The idea that economic development, usually measured as the log of per-capita gross domestic product (GDP), is related to the level of democracy in a country is highly influential in economics, sociology and political science (Lerner, 1958; Lipset, 1959; Kennedy, 2010; Inglehart and Welzel, 2009). This argument, usually labeled *modernization theory*, also has had many critics since it was first propounded almost 70 years ago (O'donnell, 1973; Robinson, 2006).

To analyze both this relationship, and some of the reasons it has been so controversial over the last sixty plus years, we will be using a data set compiled by Dr. Pippa Norris of Harvard University.

In this chapter, we will present some basic plots of this relationship that help explain both the endurance of modernization theory and why it has been so controversial. If you follow along in the exercises, you will gain additional insight.

4.2   The Data and Preliminaries

Let's start by focusing on two of the most commonly used plots in social science research: histograms and scatterplots. As always, you should start this section by opening a new session in RStudio and setting your working directory to the folder from which you will be working or open the R project, `.Rproj`, file located in the directory from which you wish to work.

```
# Set your working directory
setwd(choose.dir())
```

After you have started a new R session, load the `tidyverse` library (be sure to first install `tidyverse` using the `install.packages` command if not already done), which includes the `ggplot2` library for plotting data. We will also load our country dataset.

And let's do some simple data manipulation to get our variables into the format we would like. There are several major changes that we make. First, the original data set has almost 3,000 variables. Working with so much data can quickly become unwieldy, so we use the `select()` function to narrow the data set to just the variables we will use in the examples and exercises. Next, we use the `filter()` function to keep only the cases that have an assigned region. As a review, this is an example of combining commands. `!is.na()` is a logical test that evaluates to `TRUE` when the row is not missing a value for region and `FALSE` when it is missing (remember the `!` means `NOT`). `filter()` removes rows that do not meet the criteria in its parentheses.

Finally, we use `mutate()` to create new variables, or new versions of our variables. The first set of variables we will be using is the Freedom House measures of democracy. Freedom House is a non-governmental organization dedicated to spreading democracy worldwide. Every year since 1973, they have produced a democracy score for every country on a 7-point scale, where 7 means the country is the least free on their scale and 1 indicates a country is the most free on the scale. In this data set, the variables containing the Freedom House scores for 1972 (`fhrate72`), 1985 (`fhrate1985`), and 2008 (`fhrate2008`) are selected. But the data set saves these as a "string" because, instead of placing a 7, the data set has "Least Free," and instead of a 1, the data set has "Most Free." We convert these strings to their numeric values using the `replace()` function, convert them to numeric using the `as.numeric()` function, and, finally, subtract those values from 8 to reverse the scale – making higher numbers represent higher levels of democracy. If this discussion of string and numeric variables is a little confusing, do not worry. We will be covering these concepts in greater detail in the *Essential Programming* chapter.

For region, we use our `case_when()` function to create a new variable called "Region" that combines several categories. This will make our visualization easier later on.

For per-capita GDP, we have three measurement time periods: 1971 (`GDPPC1971`), 1984 (`GDPPC1984`), and 2007 (`GDPPC2007`). We use the `log()` function to place all three of these on a log scale, which is a common method for dealing with this particular data.

```
# Modify variables into the format we would like
ctydta_short <- ctydta %>%
  dplyr::select(Nation, fhrate72, fhrate85, fhrate08,
              Region8b, GDPPC1971, GDPPC1984, GDPPC2007,
              Fragile2006, OECD) %>%
  filter(!is.na(Region8b)) %>%
  mutate(fhrate72 = replace(fhrate72,fhrate72=="Least free","7"),
         fhrate72 = replace(fhrate72,fhrate72=="Most free","1"),
         fhrate72 = 8 - as.numeric(fhrate72),
         fhrate85 = replace(fhrate85,fhrate85=="Least free","7"),
         fhrate85 = replace(fhrate85,fhrate85=="Most free","1"),
         fhrate85 = 8 - as.numeric(fhrate85),
         fhrate14 = replace(fhrate08,fhrate08=="Least free","7"),
         fhrate14 = replace(fhrate08,fhrate08=="Most free","1"),
         fhrate14 = 8 - as.numeric(fhrate08),
         Region=case_when(Region8b=="industrial"~"Industrial",
                          Region8b=="latinameri"~"Latin America",
                          Region8b=="africa"~"Africa & M.E.",
                          Region8b=="arab state"~"Africa & M.E.",
                          Region8b=="c&eeurope"~"Eastern Europe",
                          Region8b=="se asia &"~"Asia",
                          Region8b=="south asia"~"Asia",
                          Region8b=="east asia"~"Asia"),
         ln_gdppc_71 = log(GDPPC1971),
         ln_gdppc_84 = log(GDPPC1984),
         ln_gdppc_07 = log(GDPPC2007))
```

```
## Warning in mask$eval_all_mutate(dots[[i]]): NAs introduced
## by coercion
```

Note that you will likely receive a warning message indicating that NAs were introduced by coercion. This is normal when converting a string variable to numeric, since empty strings - "" - will be automatically converted to missing values. We use the tidy-friendly `skim()` function from the `skimr` package addressed at length in the *Exploratory Data Analysis* chapter to check that everything in our resulting data looks as expected, and it does.

```
library(skimr)
```

```
skim(ctydta_short)
```

We can also leverage the `glimpse()` function from the tidyverse `tibble` package to get a printout of all features in wide format, which offers a helpful quick look at the structure of the data.

```
glimpse(ctydta_short)
```

```
## Rows: 193
## Columns: 15
## $ Nation       <chr> "Afghanistan", "Albania", "Algeria"...
## $ fhrate72     <dbl> 3.5, 1.0, 2.0, NA, NA, NA, 3.5, 2.0...
## $ fhrate85     <dbl> 1.0, 1.0, 2.0, NA, 1.0, 5.5, 6.0, 1...
## $ fhrate08     <chr> "5.5", "3", "5.5", "Most free", "5....
## $ Region8b     <chr> "arab state", "c&eeurope", "africa"...
## $ GDPPC1971    <dbl> 883.4850, 2533.3756, 3699.2895, NA,...
## $ GDPPC1984    <dbl> 1020.8060, 2994.4853, 5528.0921, NA...
## $ GDPPC2007    <dbl> 752.4724, 4729.8822, 6421.2448, NA,...
## $ Fragile2006  <chr> "Fragile", "Intermedia", "Fragile",...
## $ OECD         <chr> "Not member", "Not member", "Not me...
## $ fhrate14     <dbl> 2.5, 5.0, 2.5, NA, 2.5, 6.0, 6.0, 3...
## $ Region       <chr> "Africa & M.E.", "Eastern Europe", ...
## $ ln_gdppc_71  <dbl> 6.783874, 7.837308, 8.215896, NA, 8...
## $ ln_gdppc_84  <dbl> 6.928348, 8.004528, 8.617598, NA, 7...
## $ ln_gdppc_07  <dbl> 6.623364, 8.461656, 8.767367, NA, 8...
```

4.3 Histograms

Let's start by making a simple historgram to show the density of different democracy levels in 1972 in Figure 4.1. Histograms are a common starting place for describing our data, giving us a general idea of what the dependent variable of modernization theory looked like at the beginning of the period we are exploring. If we were doing this using base R, the commands might look something like this.

```
# Create the histogram
hist(ctydta_short$fhrate72,
     xlab = "Level of Democracy",
     ylab = "Number of Countries",
     main = "Histogram of Democracy in 1972")
```

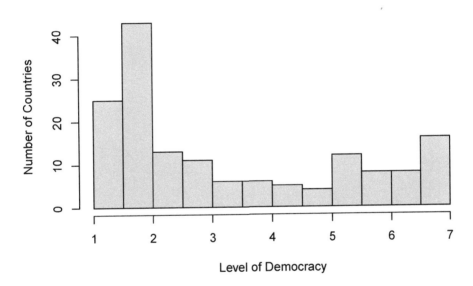

FIGURE 4.1
A Simple Histogram

There is nothing particularly wrong with this approach to plotting the histogram, but using the base R plotting functions can quickly produce clunky code that is difficult to remember, reproduce, and understand. Let's say, for example, we want to see separate histograms for the density of democracies by region: "Industrial", "Latin America", "Africa and M.E.", "Eastern Europe", and "Asia." Here is what this might look like using R's standard plots in Figure 4.2.

```
# Subset the data into regions
industrial <- subset(ctydta_short,
                     Region == "Industrial")
latin_america <- subset(ctydta_short,
                        Region == "Latin America")
africa <- subset(ctydta_short,
```

```
                        Region == "Africa & M.E.")
eastern_europe <- subset(ctydta_short,
                             Region == "Eastern Europe")
asia <- subset(ctydta_short,
                   Region == "Asia")

par(mfrow = c(3, 2)) # places histograms in 3x2 plot
hist(industrial$fhrate72,
     xlab = "Level of Democracy",
     ylab = "Number of Countries",
     main = "Industrial")
hist(latin_america$fhrate72,
     xlab = "Level of Democracy",
     ylab = "Number of Countries",
     main = "Latin America")
hist(africa$fhrate72,
     xlab = "Level of Democracy",
     ylab = "Number of Countries",
     main = "Africa & M.E.")
hist(eastern_europe$fhrate72,
     xlab = "Level of Democracy",
     ylab = "Number of Countries",
     main = "Eastern Europe")
hist(asia$fhrate72,
     xlab = "Level of Democracy",
     ylab = "Number of Countries",
     main = "Asia")
par(mfrow = c(1, 1)) # reset plot space
```

There are a number of reasons this code is less than appealing. Notice that there is no overall "grammar" to how we construct the plots. We find ourselves using $ and functions like par() that do not seem to fit with the tasks in which we have the most interest. The names we are using for the tasks also do not match easily with what we are trying to do. For example, mfrow is a vector of length 2 that specifies the number of rows and columns. But trying to remember what this command is and what it does is difficult, meaning that you will probably have to look it up next time you want to do it. There is also the issue of setting the global options. Notice that we have to use the par() function twice. The second time is to make sure we do not accidentally create a plot with 2 rows and 2 columns when we do not want to do so. It would be far better if we could revert to this default without having to do so explicitly every time. As we can attest, it is *far too easy to miss this step* and very frustrating when it happens. Finally, notice that there is no overall title

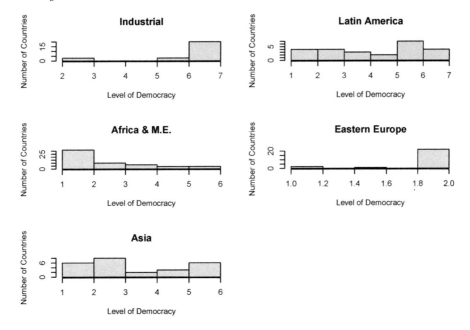

FIGURE 4.2
Separate Histograms

for the figure. This is because this is created as three separate figures pasted together.

The package `ggplot2`, included in the `tidyverse`, is designed to be a "grammar of graphics" (Wilkinson, 2012) similar to the design of `dplyr` as a "grammar of data manipulation." It has a set of commands that are consistent across different types of plots. It also, as we will see, allows you to make complex plots without a lot of extra work. Put simply, it addresses some of the problems of clunky code that has been built up over the extended period of R's initial development.

Let's make do the same thing we did above, but using `ggplot2`. Start by creating a simple histogram in Figure 4.3.[1]

```
# ggplot version
ggplot(data = ctydta_short) +
  geom_histogram(aes(x = fhrate72), binwidth = 1) +
  labs(x = "Level of Democracy",
```

---

[1]Readers should note that there exist "quick" work-arounds in `ggplot2` including both `qplot()` and `quickplot()`, which are the same. These allow for quick base-R-flavored syntax for plotting, but are more limited in certain ways than the main `ggplot()` function. Thus, mostly throughout the book we will stick with `ggplot()`, but will occasionally use one of the "quick" versions for demonstrative purposes.

```
      y = "Number of Countries",
      title = "Histogram of Democracy in 1972") +
theme_minimal()
```

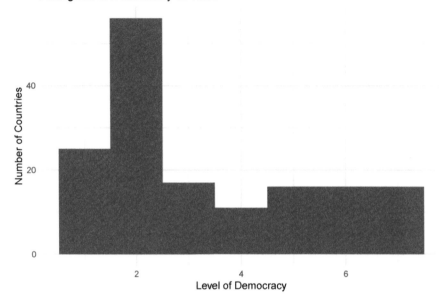

FIGURE 4.3
A Simple Histogram via ggplot

You will immediately see a few differences in this way of writing the code. First, we are combining different parts of the plot using a +. Just like the %>% we used for data manipulation, this lets us add different parts to the chart as we go.[2] To more clearly demonstrate this point, we can build a plot piece-by-piece, starting simply and adding to it to create something more complex that looks exactly the way we want it. The result is a complete plot like the one shown in Figure 4.4. Importantly, when building a plot one layer at a time like this, the use of <- allows us to save our plot as an object, and add onto that object as we go. We first save the basic histogram as an object called basic_hist, and then add to it.

```
# ggplot with additions version
basic_hist <- ggplot(data =  ctydta_short)

basic_hist <- basic_hist +
```

---

[2]In fact, we can even combine the %>% and + in ggplot2 syntax, e.g., piping the data to ggplot() and building accordingly.

```
geom_histogram(aes(x = fhrate72), binwidth = 1)

basic_hist <- basic_hist +
  labs(x = "Level of Democracy",
       y = "Number of Countries",
       title = "Histogram of Democracy in 1972")

basic_hist <- basic_hist + theme_minimal()

basic_hist
```

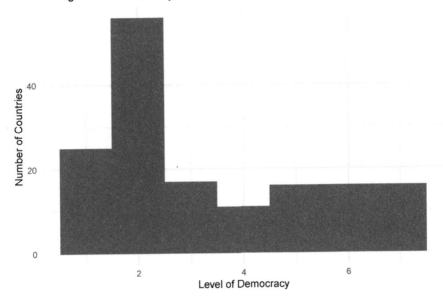

FIGURE 4.4
Building Out a Histogram

By building out the plot in this way, the grammar is more explicit. The first function is `ggplot()`, which tells R that we are using `ggplot2` to create the graph and to set up the system accordingly. In the function call, we declare the data we will be using for this plot. Since we are using only one data source for the entire plot, we can specify that data source here.

Once the system and data are declared, the next step is to tell R what *type* of chart we want to use. This involves using a "geometry" function (or "geom" for short). Here the geom is a histogram, so we will use `geom_histogram()`. As you might guess, other types of charts have similar functions: `geom_bar()` for bar charts, `geom_line()` for line graphs, `geom_point()` for scatterplots, etc. There may be some times when you want to use multiple data sources (for example, overlaying a scatterplot of one source of data with a histogram from another). To do this, simply set the data variable within the desired geometry, rather than setting it in the ggplot() function.

Within the geom, we declare a *mapping aesthetic* (or "aes" for short). The aesthetic tells the system what we want placed where. Since we are creating a histogram with the count of cases in each bin, we only need to declare our x axis (i.e. that we want the distribution of the variable fhrate72). Within the geom, we can also change a number of the options for the chart. In this case, we tell it how large we want the bins of the histogram to be (1 in this case).

There is also a function for naming the axes. We add in a set of labels for the x-axis, y-axis and the main label using the labs() function.

Finally, we change the theme from the default to a black-and-white scheme by adding the theme_minimal() function.

One thing that you will see when you run either of these code examples is a warning saying that the program Removed 36 rows containing non-finite values (stat_bin). This is simply telling you that there were 36 cases in which the variable fhrate72 was missing data, i.e., Freedom House did not provide them with ratings that year. This is another advantage to ggplot2; it tells you more about your data than the base R functions.

To see how this creates cleaner code, it is useful to show a more complex example. Let's try breaking down the histograms by region again leveraging ggplot2 and shown in Figure 4.5.

```
# Plot differences between democracy by region
ggplot(data = ctydta_short) +
  geom_histogram(aes(x = fhrate72), binwidth = 1) +
  theme_minimal() +
  facet_wrap( ~ Region, ncol = 2) +
  labs(x = "Level of Democracy",
       y = "Number of Countries",
       title = "Levels of Democracy by Region in 1972")
```

In this example, we simply added another function using the +, called facet_wrap(), which tells the system to compile subgraphs as a function of the region of the country.

And finally, ggplot2 is very flexible in allowing us to modify our charts to the particular style we want. Let's see another simple case, plotting the density of democracy scores as a histogram to see how things change when you update the bin size and colors in Figure 4.6.

```
# Update binwidth and color
ggplot(data = ctydta_short) +
```

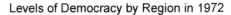

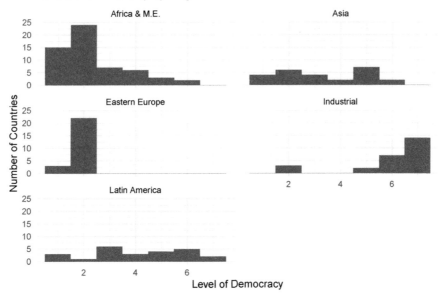

FIGURE 4.5
A More Complex, Faceted Histogram

```
geom_histogram(aes(x = fhrate72),
               binwidth = 2,
               color = "white",
               fill = "steelblue") +
theme_minimal() +
labs(x = "Level of Democracy",
     y = "Number of Countries",
     title = "Level of Democracy in 1972")
```

```
## Warning: Removed 36 rows containing non-finite values
## (stat_bin).
```

Plots created using `ggplot` are almost infinitely customizable. For those of you looking for inspiration (as well as example code), the R Graph Gallery provides hundreds of examples using the tools introduced here.

Exercises

4.3.0.0.1 Easy

- Create a histogram for the Freedom House democracy scores in 1985 and 2008. Modify the number of bins until it looks like what you want.

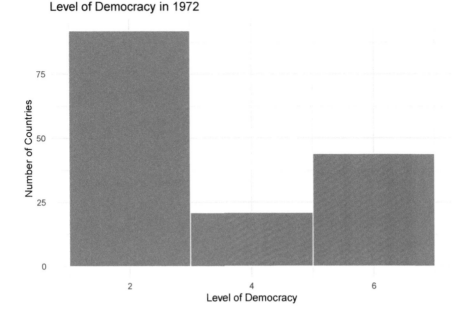

FIGURE 4.6
Updating Style and Color

What does this tell you about the progress of democracy over this time period?
- Using `facet_wrap()`, break down the democracy scores for 1985 and 2008 down by region. Be sure to set the `binwidth` and make the labels accurate. Has the progress of democracy been equally distributed across regions?
- `ggplot2` includes many different themes that can fit your personal preferences. Try changing `theme_minimal()`. Re-create the final plot for `fhrate72` from the text above, and try out `theme_bw()` and `theme_dark()` to see what some of these look like.

4.3.0.0.2   Intermediate

- In using the histogram to plot the density of `fhrate72` above, you may have noticed the warning message: *Removed 36 rows containing non-finite values (stat_bin)*. What does this mean?
- What is the difference between `bins` and `binwidth` for plotting histograms? Consider exploring the `ggplot` documentation for the answer.

4.3.0.0.3   Advanced

- This data set is not tidy, as discussed in the previous chapter. Namely, a tidy data set should have a single row for each country-year (or in this case, for every country-year pair, e.g., country observation of per-capita GDP from 1984 should be paired with the Freedom House rating from 1985). Create this tidy data set. Once you have done so, how does this change your commands to create the histogram for Freedom House scores from 1972?

4.4 Bar Plots

Another very common plot for better understanding the distribution of your data is the bar plot. Let's say you wanted to plot the number of countries in each region in your data. Since region is a categorical variable, a histogram does not make a lot of sense.

Here is how you would need to create a bar plot in base R. There are some elements of this that will likely seem strange to a beginning user. In particular, it involves combining two disparate types of commands, creating a table and then saving that table before you can plot it. See this in Figure 4.7.

```r
# First, create a table and save it as an object
region_table <- table(ctydta_short$Region)

# Create a bar chart of the table
barplot(region_table,
        xlab = "Region",
        ylab = "Number of Countries",
        main = "Distribution of Countries in Regions")
```

Now we will create the same plot using `ggplot2` with the `geom_bar()` function. Notice how we do not introduce any new concepts or steps beyond those from the process for creating a histogram. Instead, we are using the same grammar, just changing the verb (function) we use for the defining the geometry of plot. See this `ggplot()` version in Figure 4.8.

```r
ggplot(data = ctydta_short) +
  geom_bar(aes(x = Region)) +
  labs(x = "Region",
       y = "Number of Countries",
       title = "Distribution of Countries by Region") +
  theme_minimal()
```

We can also combine the skills we learned for data management earlier to create more complex plots. Let's say we want to know the average democracy score in 1972. This is easy to do using `summarize()` with our graphing functions. We can even use the pipe, `%>%`, to link them together.

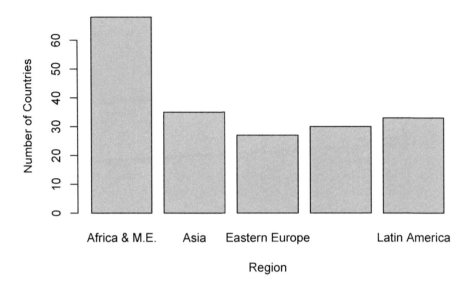

FIGURE 4.7
A Basic Barplot

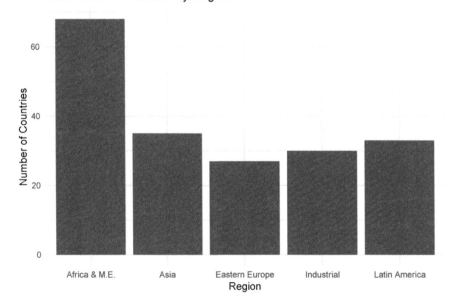

FIGURE 4.8
A Barplot via ggplot

The only new thing that we need to do is change the `stat` argument in the `geom_bar()` function to "identity". This tells the program to use the variable we define for the y-axis as the height of the bars, rather than counting the number of cases. The more complex version is now in Figure 4.9.

```
# Combine data management and visualization
ctydta_short %>%
  group_by(Region) %>%
  summarize(mean_democracy_72 = mean(fhrate72, na.rm = TRUE)) %>%
  ggplot() +
  geom_bar(aes(x=Region, y=mean_democracy_72), stat="identity") +
  labs(x = "Region",
       y = "Mean Democracy Score in 1972",
       title = "Mean Democracy Score By Region in 1972") +
  theme_minimal()
```

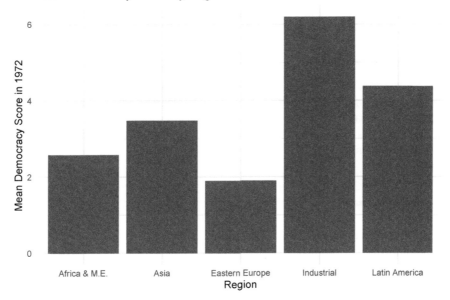

FIGURE 4.9
Complicating a Barplot with ggplot

Hopefully by this point you can appreciate how powerful the linkage between everything we have seen so far can be. The data management grammar we learned in the last chapter fits almost seamlessly with the visualization grammar we are learning in this chapter. And, in both cases, it is just a matter of putting together the right nouns (arguments) with the right verbs (functions) to convey informative statistical meaning.

Exercises

4.4.0.0.1 Easy

- Make a barplot of the stability of countries in this data set in 2006 (variable name `Fragile2006`).
- Governments that have faced political instability often struggle to subsequently democratize. Let's see if our data shows this. Create a barplot for the average Freedom House score in 2014, given stability in 2006.
- There are many options to modify bar plots. One allows you to flip the x and y axis, which is especially useful when you have long value labels for a variable. Try doing this by adding the `coord_flip()` function.

4.4.0.0.2 Intermediate

- What is the difference between a barplot and histogram? When might one be appropriate over the other?
- Update the barplot for mean democracy by region to instead summarize democracy by it's *median*. Then, render a new barplot with a different color bar for each of the five regions.

4.4.0.0.3 Advanced

- Re-produce the barplot for mean level of democracy by region, but this time using `qplot()` ("quick plot"). This is also from `ggplot2` and has some similar features, but the construction of the function, aesthetic mapping, and process of layering are all a bit different than the `ggplot()` approach we previously covered.

4.5   Scatterplots

Another very common plot in social science research is the scatterplot. This can be useful for a variety of tasks, from viewing simple distributions of variables to displaying relationships and predicted probabilities. We will discuss these exploratory-type tools in greater depth in the chapter on exploratory data analysis. As with histograms and barplots, there is a tradeoff between clunky code and less-than-exciting output versus elegant, modular code and appealing output. We start with the base R version using the `plot` command and get Figure 4.10.

```
# Scatterplot using Base R
plot(ctydta_short$ln_gdppc_71, ctydta_short$fhrate72,
    main="Relationship Between Development and Democracy, 1972",
```

```
xlab="log(per-capita GDP)",
ylab="Freedom House Level of Democracy")
```

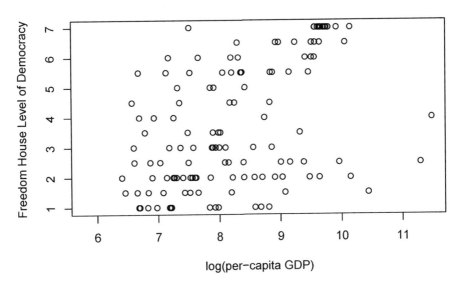

FIGURE 4.10
A Basic Scatterplot

Figure 4.10 shows a positive relationship between the log of per-capita GDP and the level of democracy for a country in 1972. Substantively, this chart shows that even in 1972, countries with a higher level of economic development measured by per-capita GDP, were generally more democratic. The relationship, however, can be a little difficult to pick out from this plot alone and there are many additional parameters that contribute to a prettier and more descriptive plot. Adding these is not very straightforward in base R.

As such, let's start with a simple `ggplot2` scatter plot with the command `ggplot()`, shown in Figure 4.11, before progressing to some more descriptive and advanced plots below.

```
# ggplot version of scatterplot
ggplot(ctydta_short, aes(x = ln_gdppc_71, y = fhrate72)) +
  geom_point() +
  geom_smooth(method = lm, alpha = 0.1) +
  labs(x="log(per-capita GDP)",
    y="Freedom House Level of Democracy",
    title="Relationship Between Development & Democracy, 1972")+
  theme_minimal()
```

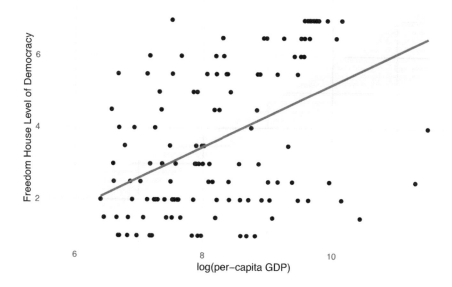

FIGURE 4.11

A Nicer Scatterplot via ggplot

This is already a much prettier plot that is also more descriptive. There are two things that we have changed from our previous plots. First, we declared our aesthetic in the `ggplot()` function, rather than in our geometry functions. Since we are using the same aesthetic for both geometries, we can declare it earlier and not have to repeat it. This is a concept that computer scientists call the "scope" of a variable. When we declare it in the `ggplot()` function, the values for the aesthetic, `aes()`, are the same for all subsequent functions and are communicated to those functions by the + operator. When we declare the aesthetic in the individual geometries, for example in `geom_point()`, it only applies to that geometry. This is very useful if we want to overlay charts using different variables, or even different data sets as mentioned above.

We also have declared two geometries, `geom_point()` and `geom_smooth()`. The first creates our scatter plot and the second creates our regression line, showing the relationship, as well as the 95% confidence intervals. The `method = lm` argument is used to specify that we are using a linear model to create the line (the default if we do not specify a method is a non-linear LOESS model).

We can do a lot with these building blocks. Let's say that we think that the effect of economic development interacts with the region in which the country is located. Indeed, some scholars have argued that the conclusions of modernization theory may be strongly influenced by geography (Ward and Gleditsch, 2018) or that there may be common historical factors leading to some countries becoming both rich and democratic, while leaving others poor and authoritarian (Robinson, 2006). We can check this by adding to the aesthetic of our plot, telling it to fill the plot components with colors representing the values of region. See the result in Figure 4.12.

```
ggplot(ctydta_short, aes(x = ln_gdppc_71, y = fhrate72,
                        color = Region)) +
  geom_point() +
  geom_smooth(method = lm, alpha = 0.1) +
  labs(x="log(per-capita GDP)",
    y="Freedom House Level of Democracy",
    title="Relationship Between Development & Democracy, 1972",
    fill="Region") +
  theme_minimal()
```

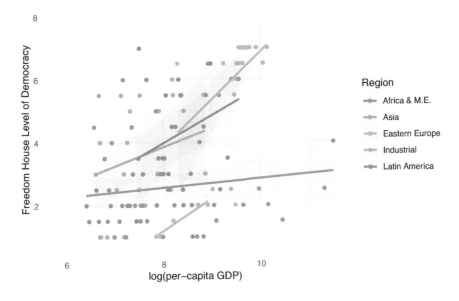

FIGURE 4.12
A Scatterplot with Continuous and Categorical Features

Note the different colors associated with the country's region. Now both the points and the linear fit lines have been colored according to the country's region, showing the relationship between economic development and democracy, conditional on region.[3] Interestingly, it looks like the relationship between economic development and democracy in 1972 is conditional on region. For industrialized, Latin American, and, within a very limited range, Eastern Euroepan countries, there seems to be a relationship between economic development and democracy. Within Asia and Africa, however, there appears to be no real relationship. Moreover, it looks like much of what we observed in the overall relationship among all countries is being driven by differences in regions. Industrialized countries are both more economically prosperous and democratic, while Asian countries at this time are both less economically developed and less democratic. It is, perhaps, not surprising that this period of time - the 1970s - was when the narrative around dependency became very popular among scholars, arguing that the world economic system was set up in such a way that the industrialized core became wealthy and democratic, while countries in the periphery remained poor and authoritarian (Smith, 1979).

There are many more arguments and updates users can make to `ggplot2` plots. For example, users can also use the `shape` argument to change the shape of the points (e.g., circles, triangles, etc.). Just run `?ggplot2` to view the many parameters and customization options available.

As we have seen before, another way to show conditional distributions is using a facet wrap, which separates each plot and places them in their own windows. The `facet_wrap()` function allows any direction or combination of columns and rows with separate plots based on the conditioning variable (input as "~ variable", where the "~" indicates that the charts are a function of the conditioning variable). The numbers of columns and rows are denoted by passing values to the `nrow` or `ncol` arguments in the `facet_wrap()` function.

Since some of the points in our plot overlap, it can be difficult to see where the largest concentrations of countries lie, so we will also set `alpha = 0.3`, which increases the transparency of points, making darker sections indicative of higher concentrations (we could also use the `geom_jitter()` function to add a small amount of noise that can make individual points more visible). The updated plot is shown in Figure 4.13.

```
# Use a facet wrap to display the regions
ggplot(ctydta_short, aes(x = ln_gdppc_71, y = fhrate72)) +
  geom_point(alpha = 0.3) +
  geom_smooth(method = lm, alpha = 0.1) +
  theme_minimal() +
  labs(x="ln(per-capita GDP)",
     y="Freedom House Level of Democracy",
     title="Relationship Between Development & Democracy, 1972")+
  facet_wrap(~ Region, ncol = 2) # update "nrow" or "ncol"
```

---

[3]We will come back to linear models and regression fit lines in the statistical modeling chapter later.

```
## `geom_smooth()` using formula 'y ~ x'
```

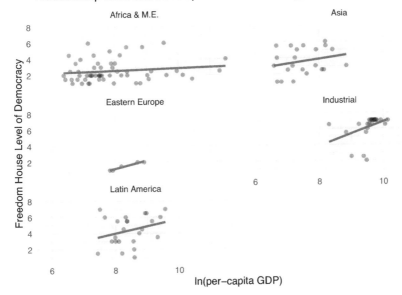

FIGURE 4.13
Faceting and Updating a Scatterplot

This set of figures tells a similar story to what we noticed above, but some of the patterns are easier to identify. We can see why modernization theory has been so controversial. While there appears to be a global pattern linking economic development and democracy, the regional heterogeneity in this relationship is such that it evokes suspicion, especially among those studying these regions.

Exercises

4.5.0.0.1 Easy

- What is the relationship between `ln_gdppc_84` and `fhrate85`? Create a scatterplot with a linear smoother to find this out.
- In the previous question, you used `method = lm` to show a linear fit line. What if we expected a non-linear relationship? Type in `?geom_smooth()` and look at the help information for the methods. What other options are available? What is the default? Try `method = loess` in one of your charts. Does this suggest a non-linear relationship?

4.5.0.0.2 Intermediate

- What assumptions are we making about the data generating process when we change the `method` argument in the `geom_smooth()` function from `lm` to `loess` to `gam` to `glm`? *Note*: We will cover data generating processes and model assumptions more in the *Essential Statistical Modeling* chapter. Thus, answer this question based on your existing knowledge.

4.5.0.0.3 Advanced

- What does the note you get in the earlier scatterplot, *geom_smooth()* *using formula 'y ~ x'*, mean? And how does `geom_smooth()` differ from `stat_smooth()`

4.6 Combining Multiple Plots

In some instances, you might want to combine multiple plots of different types. While the `facet_wrap()` function allows you to combine plots of the same type broken down by a grouping variable, it would not let you combine plots of different kinds or with different data. Unfortunately, we can't leverage the `par(mfrow)` command for `ggplot` objects as we did for plot objects in base R. As such, we demonstrate two packages for easily combining multiple ggplots: the `gridExtra` package or the `patchwork` package.

First, we will cover the use of `gridExtra`. We start by creating and storing four plots: a histogram of `fhrate72`, a bar plot for `Region`, a histogram for `ln_gdppc_71`, and a scatter plot for economic development and Freedom House democracy score. We will then paste them together using the `grid.arrange()` function and present the results in Figure 4.14.

```
# Combining plots with the gridExtra package
library(gridExtra)

plot1 <- ggplot(ctydta_short) +
  geom_histogram(aes(x = fhrate72), binwidth = 1)

plot2 <- ggplot(ctydta_short) +
  geom_bar(aes(x = Region)) +
  theme(axis.text.x = element_text(angle = 75, hjust = 1))

plot3 <- ggplot(ctydta_short) +
  geom_histogram(aes(x = ln_gdppc_71), binwidth = 2)

plot4 <- ggplot(ctydta_short) +
```

```
  geom_point(aes(x = ln_gdppc_71, y = fhrate72))
```

```
grid.arrange(plot1, plot2, plot3, plot4, ncol = 2)
```

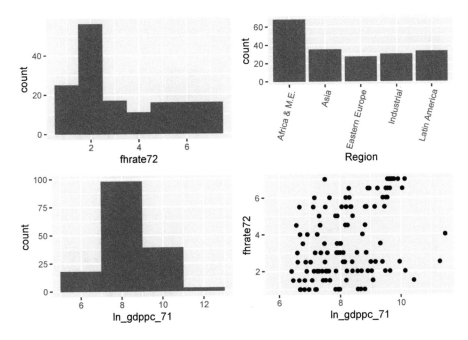

FIGURE 4.14
Combined Plots via grid.arrange

Now, let's use the same four plots in the previous case using the patchwork package. Results are in Figure 4.15.

```
# Combining plots with the patchwork package
library(patchwork)

plot1 <- ggplot(ctydta_short) +
  geom_histogram(aes(x = fhrate72), binwidth = 1)

plot2 <- ggplot(ctydta_short) +
  geom_bar(aes(x = Region)) +
  theme(axis.text.x = element_text(angle = 75, hjust = 1))

plot3 <- ggplot(ctydta_short) +
  geom_histogram(aes(x = ln_gdppc_71), binwidth = 2)

plot4 <- ggplot(ctydta_short) +
```

```
  geom_point(aes(x = ln_gdppc_71, y = fhrate72))

plot1 +
  plot2 +
  plot3 +
  plot4
```

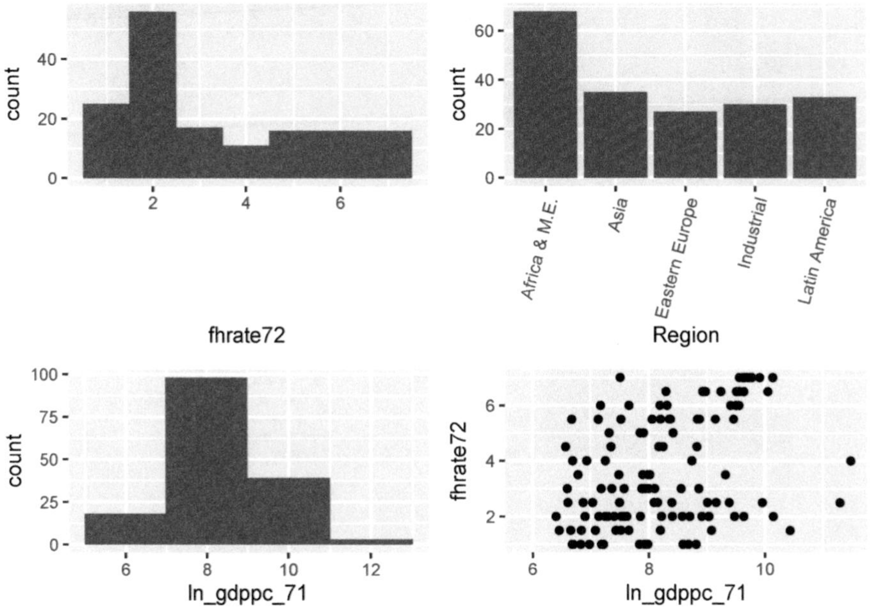

FIGURE 4.15
Combined Plots via patchwork

Though the results look pretty much the same, the `patchwork` package
offers much more flexibility in both placement of plots (e.g., using \ for
top/bottom placement), as well as in annotating the plots (e.g., using the
`plot_annotation()` function to add titles, subtitles, captions, and more). Let's
see these differences in action, all of which lead us to prefer the `patchwork`
solution over the `gridExtra` solution. See the customized result in Figure 4.16.

```
# Combining plots with the patchwork package
library(patchwork)

plot1 <- ggplot(ctydta_short) +
  geom_histogram(aes(x = fhrate72), binwidth = 1)

plot2 <- ggplot(ctydta_short) +
  geom_bar(aes(x = Region)) +
```

```
theme(axis.text.x = element_text(angle = 75,
                                 vjust = 0.5))

plot3 <- ggplot(ctydta_short) +
  geom_histogram(aes(x = ln_gdppc_71), binwidth = 2)

plot4 <- ggplot(ctydta_short) +
  geom_point(aes(x = ln_gdppc_71, y = fhrate72))

four_plots <- plot2 /
  (plot1 + plot3) /
  plot4

four_plots + plot_annotation(
  title = "Four slick plots with annotation",
  subtitle = "Here is a great subtitle!")
```

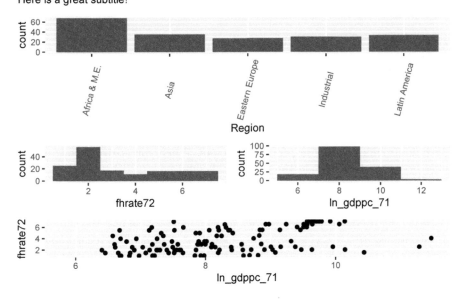

FIGURE 4.16
Customization via patchwork

Much more detail on customizing layouts, combinations of ggplot objects, and annotation options including tagging and customizing plot labels is available at the patchwork pkgdown site.

Exercises

4.6.0.0.1   Easy

- Change the layout using the + and / operators using the `patchwork` package. Create a few new layouts of the plots we previously created.
- Create four *new* plots from this data. Place them into a grid using `gridExtra`. Then do the same with `patchwork`. Can you make them look the same?
- What happens if you do not set the number of columns in the `grid.arrange()` function?

4.6.0.0.2   Intermediate

- You can create grids of multiple grids. Try `grid.arrange(g1, g1)`. What happens? What about `grid.arrange(g1, plot1)`?
- Manually add a title and subtitle using the `gridExtra` solution (*Note*: this is not as straightforward as with `patchwork`).

4.7   Saving Your Plots

Once you have spent so much time creating and cleaning your plots, it would be good if we could automatically save the plot. That way, if you need to modify it later, you can make the modifications and save them to a specified folder without needing to go through the process of finding folders or remembering how you set up the figure.

RStudio provides some tools for saving plots. When you create a plot, it will show up in the Plots tab in the lower-right-hand part of RStudio. From here, you can click on the Export menu and choose how to save your plot. You can also preview how your plot will look in different sizes.

Once you figure out what size you want your figure to be and the type of file you would like to save, you can use the `ggsave()` function to record how you want to save the figure. Let's say you want to save plot1 from the last section as a 5x7 .png file (a standard format for Microsoft Word) in a subfolder called "Figures" (make sure you have created this folder in your working directory). Here is how you can do it using `ggsave()`.

```
# Save plot1 as a .png file
ggsave(here("Figures","plot1.png"),
        plot = plot1,
        device = "png",
        width = 7, height = 5)
```

Exercises

4.7.0.0.1 Easy

- Try to save another plot to the Figures folder.

4.7.0.0.2 Intermediate

- What happens if you remove "Figures" from the `here()` function? Where does this save?
- How would you save to a particular folder without using the `here()` function? (Try to do this with a command, rather than using the RStudio dropdown menu.)

4.8 Advanced Visualizations

So far, we have created nice, clean visual descriptions of our data, but we have barely scratched the surface of what is possible in `ggplot2`, and we will return to some additional plots in the *Exploratory Data Analysis* chapter.

Before we conclude this chapter, we want to provide a few examples of advanced visualization techniques – charts that you might be less likely to use, but which demonstrate the range of R for producing visualizations. Hopefully demonstrating the range of possibilities in R will help inspire you to create your own stunning graphics.

The goal at this point, is to take that which you have learned in the comparative cases presented above, and apply it to the more complex code. Importantly, everything covered below is built using the *same* logic and syntax covered to this point.

4.8.1 Bubble Plots

In 2010, Hans Rosling presented a series of data plots to show changes in global population and health over time in a documentary on BBC Four, and in a series of YouTube videos and TED talks. These videos became incredibly popular, with just one of his videos on YouTube receiving more than 9 million views (as of this writing).

Rosling made use of animation and several other techniques which we will not cover (but, for which, there are libraries in R), but the core of his presentations was a type of chart called a bubble plot. These plots combine color and size with traditional scatter plots to present four dimensions of data in two dimensions.

The code below, resulting in Figure 4.17, demonstrates how, by simply specifying the size and color aesthetics, we can create a plot that shows how the level of democracy in 2014 is related to a country's change in per-capita GDP from 1971 to 2007, region, and level of democracy in 1972.

```
ctydta_short %>%
  mutate(change_gdppc = ln_gdppc_07 - ln_gdppc_71) %>%
  ggplot(aes(x = fhrate72, y = fhrate14)) +
  geom_point(aes(size = change_gdppc, color = Region),
             position = "jitter") +
  geom_abline(intercept = 0, slope = 1) +
  scale_y_continuous(limits = c(0.5, 7.5, 1)) +
  scale_x_continuous(limits = c(0.5, 7.5, 1)) +
  labs(x = "Democracy Score 1972", y = "Democracy Score 2014",
       size = "Per-capita GDP Growth") +
  theme_minimal()
```

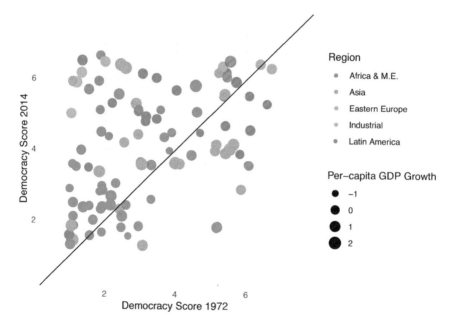

FIGURE 4.17
A Bubble Plot

There are a lot of options that have been specified in this plot, but, by now, you can probably figure out what all of them do on your own. We started by creating a new variable that records the change in the log of per-capita GDP using the `mutate` function from the last chapter.

We fed the resulting data into `ggplot` with an aesthetic that places a country's 1972 Freedom House score on the x-axis and their 2014 score on the y-axis. We then added a scatterplot using the `geom_point()` geometry, specifying that the color was determined by the country's region and the point's size was determined by the country's GDP growth. The position of the points is also specified as "jitter", which adds a small amount of noise to the data so the points that overlap are not hidden.

Since we are looking at the change in democracy, it is useful to have a line which indicates no change – in this case a 45 degree line – which we add by creating with the `geom_abline()` geometry and specifying a line with a y-intercept at zero and a slope of 1. The points above this line are the countries that have increased their democracy scores, and the ones below this line have decreased their scores.

Finally, we clean up the chart by specifying the axis marks on the x-axis and y-axis using the `scale_x_continuous()` and `scale_y_continuous()` options with the limits defined – in this case, we had the limits go from 0.5 to 7.5 by steps of 1. And, as before, we specify informative axis and legend labels, and change the theme to a black-and-white theme.

The results are somewhat more basic than Rosling's famous plots, but not by much. As you become more skilled with graphics, you will find that the sky is the limit in creating your graphics.

4.8.2   Interactive Plots

In contemporary scholarship, an increasing number of scholars are posting graphics online that can convey more information, through animation and interaction, than on the printed page. These graphics allow readers to get a better understanding of the data, as well as providing interesting summaries of data that can be shared on social media to reach a broader audience. In this subsection, we will demonstrate how to generate interactive plots.

There are many ways to generate interactive plots in R, such as `iplot`, `Rggobi`, `plotly`, and so on. Though all of these have their strengths and weaknesses, we will focus on `plotly`, which leverages `ggplot2` and is a rapidly developing platform for complex and impressive interactive plots. Further, plotly has a host website that allows users to place interactive plots on the web, and also offers a simple point-and-click interface at https://plot.ly/.

For interactive plots using `plotly` and `ggplot2`, we will need to install and load the `plotly` R package. Also, in the code below we will be returning to the pipe operator, `%>%`, which is a centerpiece of the Tidyverse. We will continue to use our `ctydta_short` data object. For these plots, our key function is `plot_ly`, where the argument `type` allows us to change the plot type (e.g., scatterplot, histogram, etc.). Run the following code locally for a simple interactive scatterplot.

```
library(plotly)

# simple scatterplot
scatter <- plot_ly(ctydta_short,
                    x = ~ ln_gdppc_71,
                    y = ~ fhrate72,
                    type = "scatter", # plot type
                    text = paste("Country: ",
                                    ctydta_short$Nation), # hover
                    mode = "markers", # object type
                    color = ~ Region,
                    size = ~ ln_gdppc_71
) %>%
  layout(title='Simple Scatterplot',
      xaxis=list(title='ln(per-capita GDP)'),
      yaxis=list(title='Freedom House Democracy Score, 1972'))
scatter
```

The plot you see in the Viewer tab in the lower right-hand corner of RStudio allows you to hover over the points to see more information. You can save this plot as an interactive HTML plot by going to the Export dropdown menu and selecting "Save as Web Page...". (Note that, if it does not show up in the Viewer window, this is likely just a result of the size of your screen. You can save it as a web page and it will show up when you click the saved file.)

In the code chunk above, there are a few changes to syntax, but mostly the intuition from ggplot2 remains the same. Note, though, that for axis titles, we need to pipe in a new layer (layout). Further, take note of the fact that we stored the plot in its own object, scatter, which is a requirement for building interactive plots manually like this in R.[4]

Let's try another interactive plot, but of a histogram.

```
# Interactive histogram
hist <- plot_ly(ctydta_short,
                x = ~ fhrate72,
                type = "histogram",
                text = paste("Region: ", ctydta_short$Region),
                color = ~ Region
) %>%
  layout(title='Simple Histogram',
      xaxis=list(title='Freedom House Democracy Score, 1972'),
      yaxis=list(title='Number of Countries'))
hist
```

---

[4]This is not necessary for building plots directly on Plotly's website.

There is a ton of information in this plot! Inspect it all carefully. Try leveraging the internal tools from `plotly` in the upper right, such as changing the number of bars viewed at a time, the hover text, magnifying portions, and so on. These are quite informative, powerful plots that do not require a great deal of additional knowledge beyond our use of `ggplot()` above.

4.9 Concluding Remarks

This has been a whirlwind tour of visualization using `ggplot2` and other related packages in the broader Tidyverse. While we have covered a lot of ground, the reality is that we have only scratched the surface of the visualization tools available and what you can do with these tools. As you have probably figured out by now, R is incredibly flexible for visualization, allowing for a wide range of plot options that are basically impossible in statistical packages like SPSS, SAS, and Stata.

If this sounds like the area of artists, you are not far off. While many social scientists fall into the trap of re-creating the basic plots they learn in their introductory classes over and over, truly great visualizations find ways to convey information in a manner that is beautiful and meaningful (Tufte et al., 1990; Tufte, 2001; Healy, 2018). You now have the fundamentals needed to produce such plots, so continue learning visualization and let your imagination run free.

4.9.1 More Resources

There are several nice resources that you can access online or in print to help you with more specific plots. Here are two excellent resources to help you as you continue to develop your skills with ggplot.

1. The Data Visualization Chapter in the *R for Data Science*, which is a free book by Hadley Wickham and Garrett Grolemund (Wickham and Grolemund, 2017). A version of this is available free online.

2. For a book length treatment of the subject, Kieran Healy's book, *Data Visualization: A Practical Introduction* (Healy, 2018), is an excellent treatment of visualization generally, and visualization using `ggplot2` in particular.

5

Essential Programming

In this chapter we are depart from the Tidyverse for a while to introduce readers to core programming concepts in (mostly, base) R. These tools are invaluable for efficiently engaging with R programming, both in and out of the Tidyverse. Though the majority of coverage in this chapter is using base R tools, at the end we will return to the Tidyverse, covering a core functional programming task – *mapping* (via the Tidyverse `purrr` package). Our goal here is to cover a variety of tools and syntactic choices in R to widen *and* deepen your R toolbox, driving toward the ultimate goal of making our way up the steep R learning curve.

This chapter is a bit more technical than applied, but is no less important for cultivating an effective understanding of R. In the long term, understanding the basics of programming in R will help to expand your horizons and open up new vistas for your research.

5.1 Data Classes

Before we get into the ins-and-outs of programming, the next couple of sections will look under the hood of R to discuss some of the fundamental items that make up the language. One way to think about this is that we will be looking at the small building blocks that can be used to make a much larger structure.

We start by discussing the types of data objects R allows you to use and how they behave. There are several different classes of data, and the operations you can perform on the data will differ, depending on the class. These classes are "numeric", "character", and "factor."

Numeric data is just what it sounds like. This is data that is either made up of integers or doubles (a term used for numeric data that can have decimal places).

```
# Examples of numeric class
class(1)
```

```
## [1] "numeric"
```

```
class(1020)
```

[1] "numeric"

```
class(0.50)
```

[1] "numeric"

Character data (sometimes called "strings") is data made up of a combination of letters, numbers, and, sometimes, symbols. Character data will be encased in quotation marks when it is printed out.

```
# Examples of character class
class("this is a string")
```

[1] "character"

```
class("email@email.com")
```

[1] "character"

```
class("1")
```

[1] "character"

Notice especially the last example. It is the number 1, so you might expect it to be numeric, but it is not because it is in quotation marks. If you were to try a numeric operation, say adding it to something else, R would give you an error.

```
# Adding a string with a numeric
"1" + 1
```

Finally, factor data is one of the most confusing classes to deal with in R, and many people choose to avoid using it unless it is specifically needed. The factor class is a hybrid between the numeric class and the character class. For most functional purposes, it is treated as being of the character class, but with an underlying order. Here is an example of how this can get confusing.

```
# Example of factor class
example <- as.factor(c("1","3","5","2","5","1","100"))
example
```

```
## [1] 1    3    5    2    5    1    100
## Levels: 1 100 2 3 5
```

This tells us that we have created a vector of factors with 5 levels (representing the unique strings in the vector). Now let's say we look at the numbers in this vector and say we want to treat them as numbers.

```
# Example of problems with factor class
```

```
strange_example <- as.numeric(example)
strange_example
```

[1] 1 4 5 3 5 1 2

This is not at all what we expected. How did "100" become 2? And how did "3" become 4? This is because it is giving you the underlying numbers behind the factors, not the values of a numeric version of the strings. To do this, we would need to do something like this.

```
# Convert a factor class to numeric
expected_example <- as.numeric(as.character(example))
expected_example
```

[1]   1   3   5   2   5   1 100

And this is one of the reasons we will avoid factors for most of our work in this book. There are, however, some situations in which factors are useful. For example, factors can be used to set an order to character data for graphing. If you want to create a factor variable, it is recommended that you also explicitly set the order of that variable using the `levels =` option. In the following example, we want low to be associated with 1, medium with 2, and high with 3.

```
# Create a set of values
values <- c("high", "low", "medium", "low", "low", "high")

# Create an unordered factor
unordered_factor <- factor(values)

# Create ordered factor
ordered_factor <- factor(values,
                         levels = c("low", "medium", "high"))

# Print them both out for comparison
unordered_factor
```

```
## [1] high    low     medium low     low     high
## Levels: high low medium
```

```
ordered_factor
```

```
## [1] high    low     medium low     low     high
## Levels: low medium high
```

```
# Show them both as numeric
as.numeric(unordered_factor)
```

[1] 1 2 3 2 2 1

```
as.numeric(ordered_factor)
```

```
## [1] 3 1 2 1 1 3
```

As you can see above, you can move between data types, to a degree. The character "1" can be made numeric using the `as.numeric()`, or a set of strings can be converted into factors using `factor()`. You should always be careful when you move between classes, however, to make sure you do not get unexpected results.

Exercises

5.1.0.0.1 Easy

- Classify each of the following as numeric or character: 2, "two", "five", "5", 100.
- Extending the logic we've covered so far, what is the function you would use to check whether an object is a matrix?

5.1.0.0.2 Intermediate

- As shown above, if you try entering `"1"` + 1, you will get an error. Create two variables, one called `number_one` with a value of 1 and one called `character_one` with a value of `"1"`. How can you make `number_one` + `character_one` produce the correct answer, 2?

5.1.0.0.3 Advanced

- Create a vector of values using `numeric_vector <- runif(10)`, which gives you 10 random numbers between 0 and 1 from a uniform distribution. What happens when you use `as.character()` on this vector? What happens when you use `as.factor()`? What happens when you use `as.numeric(as.factor())`? Why?
- Describe the differences between factor objects and categorical objects.

5.2   Data Structures

There are several main data structures used in R, some of which we have already encountered. The first, a vector, we have seen before. A vector is a one-dimensional collection of objects and can hold a set of items of any class. However, it cannot hold items of different classes. If it receives inputs from different classes, it will change these to make them a single class.[1]

---

[1]In most other programming languages, a vector is called an "array". Technically, an "array" in R is a separate data structure, which can be comprised of vectors of one or more

```
### A vector
vector1 <- c(1,2,3,4,5) # numeric
vector1
```

```
## [1] 1 2 3 4 5
vector2 <- c("jack", "jill", "up the hill") # character
vector2
```

```
## [1] "jack"      "jill"        "up the hill"
vector3 <- c(1,2,"three","four") # num + char --> char
vector3
```

```
## [1] "1"    "2"      "three" "four"
```

Because they only include items of the same class, we can check the class of items in a vector. As you can see below, if you include both character and numeric items in a vector, the numeric entries are converted to character entries and the vector will be considered of the character class.

```
# Class of items in a vector
class(vector3)
```

```
## [1] "character"
```

You can access single items in a vector by placing the index of the item you want in brackets. If you wish to access multiple items, you can specify the range using a colon (:). This will return the items in the vector from the first number to the second, inclusive of both.

```
# You can access items in a vector using vectorname[#]
## e.g., to access the 2nd item in vector1
vector1[2]
```

```
## [1] 2
## Access the 3rd item in vector2
vector2[3]
```

```
## [1] "up the hill"
## Access multiple items in a vector using `:`
vector1[2:4]
```

```
## [1] 2 3 4
```

You can also find the length of a vector using the `length()` function. This

---

dimensions. An array of one dimension is almost the same as a vector and an array of two dimensions is almost the same as a matrix. We will not go into depth on arrays, since they are not as commonly encountered as other data structures.

can be an especially useful function if, for example, you are looking at vectors of differing lengths and want to get the last few items in them. This will be important when you try to write functions that might be used on vectors of different lengths. The code block below shows an example where we get the last three items in "vector 1" by asking for all the items ranging from two places before the end of the vector to the end of the vector.

```
# Number of items are in the vector using length()
length(vector1)
```

```
## [1] 5
```

```
# Get last three items of vector1
vector1[(length(vector1) - 2):length(vector1)]
```

```
## [1] 3 4 5
```

A matrix is essentially a collection of one-dimensional vectors arranged into two dimensions of rows and columns. As with a vector, a matrix can only be of one class. And, while we will not cover it here, R includes a range of operations that can be used for matrix (linear) algebra. You can access parts of a matrix by placing the numbers of the rows and columns you want into brackets, separated by commas. The first item in the brackets is the row number, the second is the column number. So, for example, `matrix1[2,3]` requests the entry in the second row and the third column of the matrix.

```
matrix1 <- matrix(c(1,2,3,4,5,6,7,8,9), nrow = 3, ncol = 3)
matrix1
```

```
##      [,1] [,2] [,3]
## [1,]    1    4    7
## [2,]    2    5    8
## [3,]    3    6    9
```

```
# You can access items in a matrix using matrixname[#row, #col]
## e.g., access the value in the 2nd row of the 3rd column
matrix1[2,3]
```

```
## [1] 8
```

```
## Use `:` to access multiple; and blank to access all items
matrix1[2:3,]
```

```
##      [,1] [,2] [,3]
## [1,]    2    5    8
## [2,]    3    6    9
```

```
## Find the dimensions of your matrix using dim()
dim(matrix1)
```

```
## [1] 3 3
```

```
## Find the number of rows and columns with nrow() and ncol()
nrow(matrix1)
```

```
## [1] 3
```

```
ncol(matrix1)
```

```
## [1] 3
```

A list is a collection of other data structures (including lists of lists). Unlike vectors and matrices, a list can contain any combination of data types. Lists are especially useful for programming because they give you a very flexible data structure for storing a range of values. In the example below, we show you the creation of a list that includes two vectors and a matrix. You can also see different methods for accessing the items in a list. One way is to use double brackets ([[ ]]). So, for example, list1[[2]] gets the second element in the list – in this case it is "vector2". Another way is to assign names to the list items. In the example below, we assign the names using the **names()** function on the left and assigning a vector of names on the right. Once a list has been assigned names, you can access the items in the list using a $, where the list name is on the left side and the name of the list item you wish to access is on the right.

```
# Lists are a collection of other data structures
## They are what most statistical functions return
list1 <- list(vector1, vector2, matrix1)

# Items in a list can be accessed using listname[[itemnumber]]
list1[[2]]
```

```
## [1] "jack"        "jill"        "up the hill"
```

```
# Name and access items in the list via listname$itemname
names(list1) <- c("vector1", "vector2", "matrix1")
list1$vector2
```

```
## [1] "jack"        "jill"        "up the hill"
```

In fact, if you have used R before, you were probably using lists without even knowing it. Dataframes and tibbles are both special kinds of lists. Indeed, you have seen us use the $ before to access particular variables in a dataframe or tibble. Similarly, when you run a regression model, the object that is returned is a list with a range of different components like the coefficients, the variance-covariance matrix, etc.

In the example below, we run a regression of y1 on x1 (both random numbers from a uniform distribution). When we use the **names()** function, you can see that, as in the list example above, we get names for the items in the resulting list. One of those items is a vector of coefficients, which we can access using the $. We can also access specific items in the vector using the brackets ([ ]), just like we observed above. Finally, we can use the list in the summary() function to see the results in a nice format. If you repeat the process on the results of the summary() function, you will see that the output of that function is also a list.[2]

```
# Using lists in R without knowing it
y1 <- runif(100, min = 0, max = 1)
x1 <- runif(100, min = 10, max = 20)
regression <- lm(y1 ~ x1)
names(regression) # list returned by lm()
```

```
##  [1] "coefficients"  "residuals"       "effects"
##  [4] "rank"          "fitted.values"   "assign"
##  [7] "qr"            "df.residual"     "xlevels"
## [10] "call"          "terms"           "model"
```

```
regression$coefficients # coefficient vector in lm()'s return
```

```
## (Intercept)          x1
##   0.6446093  -0.0094899
```

```
regression$coefficients[2] # Accessing one of the coefficients
```

```
##          x1
## -0.0094899
```

```
summary(regression) # summary == presentation of list elements
```

```
##
## Call:
## lm(formula = y1 ~ x1)
##
## Residuals:
##      Min       1Q   Median       3Q      Max
## -0.53854 -0.22524  0.01843  0.28101  0.52222
##
## Coefficients:
##              Estimate Std. Error t value Pr(>|t|)
## (Intercept)  0.64461    0.16239    3.97 0.000137 ***
## x1          -0.00949    0.01117   -0.85 0.397502
## ---
## Signif. codes:
```

---

[2]Readers should note that here and anywhere in the book where simulations or random draws are used, results will be slightly different from our own given the randomness of the process, e.g., runif().

```
## 0 '***' 0.001 '**' 0.01 '*' 0.05 '.' 0.1 ' ' 1
##
## Residual standard error: 0.2987 on 98 degrees of freedom
## Multiple R-squared:  0.007315,Adjusted R-squared:  -0.002814
## F-statistic: 0.7222 on 1 and 98 DF,  p-value: 0.3975
```

As mentioned above, a dataframe (data.frame) is a type of list, but it has some special features that are worth mentioning. A dataframe is a list that creates rectangular data. When you import a data set from .xls, .csv, .dat, or .dta, you get a dataframe. While it looks a lot like a matrix, it allows you to access columns by names and combine a range of data types. All of the individual columns will be of the same data type, but different columns can have different data types. You can also access individual columns using a $, much like a list. Within each column, all of the items are of the same class.

```
# Dataframes are what we think of as a data set in Stata or SPSS

## convert a matrix to a dataframe using data.frame()
dataframe1 <- data.frame(matrix1)
names(dataframe1) <- c("y", "x1", "x2") # Name the variables

# Access different variables via dataframename$variablename
dataframe1$y # Returns the variable as a vector
```

```
## [1] 1 2 3
```

```
dataframe1$y[2]
```

```
## [1] 2
```

```
# Use the head() function to see the first few rows
head(dataframe1)
```

```
##   y x1 x2
## 1 1  4  7
## 2 2  5  8
## 3 3  6  9
```

The last data structure we will discuss is a relatively new one developed for the Tidyverse called a "tibble." A tibble is very similar to a dataframe, and even can be constructed using very similar commands. Tibbles have a few advantages in speed with larger datasets and have a nicer print option (no need to use head()). Because of this, tibbles are slowly becoming the standard.

```
# Create a tibble out of a matrix
tibble1 <- as_data_frame(dataframe1)
tibble2 <- as_tibble(dataframe1)

# Print the data
```

```
tibble1
tibble2
```

Exercises

5.2.0.0.1 Easy

- Using `dataframe1`, which you created in this section, display the first 10 rows and 10 columns. How do you do this?
- When you type in `tibble2` from above, you will get a printout of the data structure. What do the letters (e.g. `STR`) under the column names stand for? Look up what the different column names can be and what they mean.

5.2.0.0.2 Intermediate

- Create two vectors, `v1 <- c(1, 2, 3)` and `v2 <- c(4, 5, 6)`. What happens when you use `c(v1, v2)`? Why? What about `rbind(v1, v2)` or `cbind(v1, v2)`?

5.2.0.0.3 Advanced

- Print the *structure* of `tibble1` and `tibble2` previously created. Describe the differences in these types of objects in substantive terms. And further, why might one use a tibble instead of a data frame?

5.3   Operators

Programming in R is built on expressions, operators, and characters. And further, when using R, we are often concerned with accomplishing complex tasks (or even simple ones) most efficiently and quickly. This implies some degree of iterating over a series of simpler tasks. While our goal is to encourage creation of user-defined functions and loops whenever possible, at a minimum, this chapter is concerned with getting you comfortable with the general syntax that is central to programming in R.

First, consider *relational* operators. These are symbols, or "operators" that specify relationships between objects. And recall that R is built on "object-oriented programming", where values are stored in objects which can be manipulated and combined a variety of ways downstream. The main relational operators are:

1. `<` less than
2. `>` greater than

3. `<=` less than or equal to
4. `>=` greater than or equal to
5. `==` equal (identical) to
6. `!=` not equal to

Relational operators return an object of class "logical", meaning it has a value of either **TRUE** or **FALSE**. Put in terms of an example, the first line of the code block below asks R whether 5 is greater than 4. R returns **TRUE**. The second line asks if 5 is less than or equal to 4. R returns **FALSE**. The third line shows that the result is of class "logical."

```
5 > 4
```

```
## [1] TRUE
```

```
5 <= 4
```

```
## [1] FALSE
```

```
class(5 > 4)
```

```
## [1] "logical"
```

Similar to relational operators are *logical* operators. These provide the ground rules for combining and pairing objects in a variety of manners. Consider the most common logical operators:

1. `!` not
2. `&` and
3. `|` or

Note that the `!` operator appears in both lists of operators. This is because, on its own it just means "not," which is a logical expression. Combined with other operators, `!` can add its value, so to speak, to others (e.g., not equal to is `!=`).

These logical operators can be used to produce complex conditions. For example, the first line of the code block below tests whether 5 is greater than 3 AND whether 5 is greater than 6. This returns a value of **FALSE** because one of the two conditions is false. The second line tests whether five is greater than 3 OR whether 5 is greater than 6. This evaluates to **TRUE** because one of the two conditions is true. The third line uses the `!` operator to reverse the second line. This evaluates to **FALSE**, reversing the result of the second line.

```
5 > 3 & 5 > 6
```

```
## [1] FALSE
```

```
5 > 3 | 5 > 6
```

```
## [1] TRUE
!(5 > 3 | 5 > 6)

## [1] FALSE
```

The last line of the above block also shows an important point about the use of () with logical operators. Just like mathematical equations, the statement inside the parentheses is evaluated first, followed by the statement outside the parentheses.

Exercises

5.3.0.0.1 *Easy*

- What does the following statement evaluate to – !((5 > 3 & 5 > 6) | 5 > 6)? Why?

5.3.0.0.2 *Intermediate*

- Think about the use of == compared to = seen elsewhere in the book. Apply this logic to the ! operator, and offer a definition of !! (read: "double bang" or "bang bang").

5.3.0.0.3 *Advanced*

- As we have discussed previously, objects that take other objects as an input and then output another object are called functions. Demonstrate that relational and logical operators are *also* functions. Try \>`(5,3)`. What does this produce? Can you produce the same statement as in #1 above using the function form?

5.4   Conditional Logic

if and if else are essential building blocks to programming in R, from testing certain values or expressions to writing packages and big chunks of code with conditional statements. They are very powerful tools in programming, and similar versions exist in all major programming languages. Specifically, the syntax starts with if, and then a value to be tested is supplied in parentheses, followed by braces, which include the statement to be expressed. In if else cases, the user can evaluate a statement under different constraints (e.g., "If value X is Y, then do Z. Otherwise ("else"), do A.").

Let's begin with a simple case of an if statement: evaluating whether a supplied number is positive, and printing as much if it is.

```
x <- 5

if (x > 0) {
  print("Positive number")
}
```

[1] "Positive number"

Next, let's take a simple case for `if else`. In this case, we check if a number is greater than zero. If so, we print `Positive`, else we print `Negative or zero`.

```
x <- -5

if (x > 0) {
  print("Positive")
} else {
  print("Negative or zero")
}
```

[1] "Negative or zero"

Note, we are creating and defining an object x, which is the value being evaluated. We can redefine x, and test it again.

```
x <- 5

if (x > 0) {
  print("Positive")
} else {
  print("Negative or zero")
}
```

[1] "Positive"

Though seemingly simple, `if` and `if else` are core to understanding and applying programming in R.

Exercises

5.4.0.0.1 Easy

- Create your own spelling test. Check if a variable contains the word "antialestablishtarianism" and have it print "correct" if it is spelled correctly or "incorrect" if it is spelled incorrectly. Try out different spellings for your variable and see what it produces.

5.5   User-Defined Functions

Building on (and soon to layer) the logic of conditional statements using if and if else, we now shift to user-defined functions. These are similarly powerful programming tools that drastically streamline the programming (and research) process. They allow users to do a ton of tasks, like automating rote, redundant code and calculations. But the value of functions is mostly that they allow for consistent calculation and for simple usage in future applications. They operate on the same principle of preferring sum(2,2,2,2) in base R to the more laborious (2 + 2 + 2 + 2). Though the tradeoff may seem minimal with the simple example, the value of writing functions to streamline code and calculations will quickly become apparent.

As before, we begin with a simple example to get the intuition: squaring a value. Rather than typing: (3^2), (4^2), (5^2), and so on, a function would streamline this process significantly, prevent the likelihood of messing up the syntax if approached line by line, and also allow the user to come back to access the function in the future (as well as update for needed complexity as we will see below). The syntax is defining a new object, and then specifying the function with an argument supplied in parentheses. Then, within braces, there is a statement to be evaluated, and the result is returned as output. With that, let's make this squared value a function.

```
sq <- function(x) {
  sqn <- x^2
  return(sqn)
}
```

With the function defined by the user (hence the name), we can call the function to see if it worked properly.

```
sq(2)
```

```
## [1] 4
```

This is good news! Our function worked as expected. Feel free to try squaring any value to verify (or have fun). Now, let's complicate our example just a little bit, allowing for greater flexibility. In the following case, we are updating the function to allow for x and y values to be defined. Thus, instead of just squaring our supplied value, we are now allowing for raising any value, x, to any power, y. As such, we change the name of the original function from sq to exp.

```
exp <- function(x, y) {
  expn <- x^y
```

```
  return(expn)
}
```

With the function defined, we can now call it to see if it worked. For a simple case, raise 2 to the power of 4.

```
exp(2,4)
```

[1] 16

Now, here is another complication, but allowing for a much more descriptive (and thus useful) function. In the next case, we are printing a descriptive output using both `print()` and `paste()`, the latter of which allows us to "paste" words along with our output, which is especially useful when writing R packages.

```
exp <- function(x, y) {
  expn <- x^y
  print(paste(x,"raised to the power of", y, "is", expn))
  return(expn) # optional
}
```

And further, we can also assign "default values" in our functions, which are values you don't have to specify, but could change if you want. So, in the example below, if the user does not define a specific value for y, it will, by default, be assigned a value of 2.

Note, we are continuing to redefine our `exp` function from earlier. If you wanted to leave the original intact, you would simply need to change the object name to the left of the assignment operator, `<-`.

```
exp <- function(x, y = 2) {
  expn <- x^y
  print(paste(x,"raised to the power of", y, "is", expn))
}
```

From here, we can call a few versions of the function to see everything come together:

```
exp(3)
```

[1] "3 raised to the power of 2 is 9"

As expected, when only one value – the x value – is defined, the default value for y is used and the function squares the provided value.

Or. . .

```
exp(3,1)
```

```
## [1] "3 raised to the power of 1 is 3"
```

In this case, we have defined both an x and y value, so the default squaring is overridden and y is assigned a value of 1.

Now, let's build on what we have learned so far and create a new function that actually does something of more value. Specifically, we can write a function that calculates temperature in Celsius, given a supplied Fahrenheit value.

```
celsius <- function(f) {
  c <- ((f - 32) * 5) / 9
  return(c)
}
```

With the function defined, we can either supply individual Fahrenheit values, or a vector of Fahrenheit values; the function can handle both. Let's store a vector of Fahrenheit values in the object **fahrenheit** and test out the function (**note**: if we supply a vector of values, we should get a vector of values returned as output).

```
fahrenheit <- c(60, 65, 70, 75, 80, 85, 90, 95, 100)

celsius(fahrenheit)

## [1] 15.55556 18.33333 21.11111 23.88889 26.66667 29.44444
## [7] 32.22222 35.00000 37.77778
```

Excellent! The function worked as expected with quick calculation of a vector of Fahrenheit values to Celsius via our **celsius** user-defined function.

Exercises

5.5.0.0.1 Easy

- Take one of the number comparison if else statements from the previous section and make it a function. Make sure you understand how this works.

5.5.0.0.2 Intermediate

- Write a function to convert pounds (lbs) to kilograms (kgs) (*note:* 1 lb ≈ 0.45 kg).

5.5.0.0.3 Advanced

- Write a function and place it within another function. This can evaluate any expression you'd like (e.g., nesting power rules). More broadly, discuss the benefits of such a task. Why might you do it? When would it *not* make sense to do so?

5.5.1 Layering Statements

In addition to learning each of these parts, it is important to note that the power of these programming building blocks is that they can be layered. Notably, we can embed conditional logic previously discussed (`if` and `if else`) into user-defined functions to make them even more powerful, descriptive, and ultimately more useful.

```r
# First, write the function
pnz <- function(x) {
  if (x > 0) {
    n <- "Positive"
  }
  else if (x < 0) {
    n <- "Negative"
  }
  else {
    n <- "Zero"
  }
  return(n)
}

# Now call it for a variety of values
pnz(4)
```

```
## [1] "Positive"
```

```r
pnz(-3)
```

```
## [1] "Negative"
```

```r
pnz(0)
```

```
## [1] "Zero"
```

Note that in the combination above, we combined `if` and `else` to have an `else if` statement, which is a programmatic way of layering multiple statements in a single function. This is somewhat similar to the layering of the `ifelse()` function we used for recoding variables two chapters ago. Indeed, the logic of `if` and `else` presented here is similar, although the programming functions are more flexible in their use.

Finally, de-bugging is a key piece of writing code in R, especially when creating R packages. Specifically, we can tell a function to `stop` if something in the code is wrong/missing, or we can also print `warning` messages if something is where it should not be, but we don't want to stop the function entirely and throw an error message.

Let's put all of these pieces together that we have learned so far and replicate a function to calculate the Herfindahl-Hirschman Index (HHI), which is a measure of market concentration (often used as a proxy for competition). This is from an R package hhi, and serves as a useful case study applying all of this logic (Waggoner, 2018b). The package takes a dataframe as its input, x, along with a string, s, that identifies the variable for which HHI is calculated.

```
# Calculate Herfindahl-Hirschman Index Scores
#
# usage: hhi(x, "s")
# x Name of the data frame
# s Vector corresponding with market shares
# return: hhi A measure of market concentration
# Note: Vector of "share" values == total share of firms
# Note: 0 = perfect competition; 10,000 = perfect monopoly


hhi <- function(x, s){
  if (!is.data.frame(x)) {
    stop('"x" must be data frame\n',
         'You have provided an object of class: ', class(x)[1])
  }
  shares <- try(sum(x[ ,s]))
  if (shares < 100 | shares > 100) {
    warning('shares, "s", do not sum to 100')
  }
  d <- x[ ,s]
  if (!is.numeric(d)) {
    stop('"s" must be numeric vector\n',
         'You have provided an object of class: ', class(d)[1])
  }
  if (any(d < 0)) {
    stop('vector "s" must contain only positive values')
  }
  hhi <- sum(d^2)

  return(hhi)
}
```

With the function defined, as well as parameters defined, we can create some fake "firm" data as well as the share of the market each retains, and then calculate the competitiveness (or concentration) of this hypothetical market.

```
a <- c(1,2,3,4) # firm id
b <- c(20,30,40,10) # market share of each firm (totaling 100%)
x <- data.frame(a,b) # create data frame
```

```
hhi(x, "b")
```

```
## [1] 3000
```

5.6 Making Your Code Modular

Once you have created a function, you can start to make your code "modular." This means that you can start to split your code between several files. Why would you want to do this?

As you conduct more and more actions during your analysis, you will find that your scripts may become quite long. We have had scripts that have run into thousands of lines. Think about it. By the time you have made all of the changes you want to a data set, done some exploration, finished your main analyses, and conducted some robustness checks, you might end up with a very long list of commands and it might be difficult to find and modify specific parts.

Also, with what you have learned in this chapter, you might want to use your functions multiple times. Take the function for calculating HHI in the last section. This may be something you will want to do in several different projects, and it can become messy to paste it into every script you write in its entirety.

So, instead of putting everything into a single file, you can save them as separate files (or "modules") and load them into your code. For the `hhi` function you created in the last section, you can put it into a script file and save it as `hhi.r` in your working directory. Once you have done this, you can load the function into another project by simply typing `source("hhi.r")`. You do not have to limit yourself to one function per module. Modules can contain any number of functions. So you could make a collection of calculations you find yourself using often and load all of those functions using this method.

If this looks a little familiar, it is basically the same thing you have been doing when you install and load a package. Packages are simply collections of functions that have some additional attributes (like the help documentation) that make them easier to use. At some point, you might want to turn some of your modules into packages. There are many resources online to help you with this, as well as Hadley Wickham's book on the subject.

In addition to saving and reusing functions, you can place entire parts of your analysis in different modules to make it easier to keep track of your analysis. If, for example, you have a hundred lines of code to take some raw data and convert it into the format you want, this can also be saved as a function and saved as a separate module.

By saving parts of your code as modules, you will make it easier to change and maintain your code.

Exercises

5.6.0.0.1   Easy

- Try this out. Create a module called `hhi.r`. Open a new script and load the module. Then try out the `hhi()` function in that new script.

5.6.0.0.2   Intermediate

- Let's say you want to have a function you can use to convert a number of imperial measures to their equivalent metric measures. Create a function called `convert` that takes two inputs – a numeric value and the name of the type of measure ("foot", "yard", and "mile"). Write the function such that if a value in feet is entered, it will convert to centimeters (1 foot ≈ 30.48 centimeters); if a value in yards is entered, it will convert to meters (1 yard ≈ 0.91 meters), and if a value in miles is entered, it will convert to kilometers (1 mile ≈ 1.61 kilometers).

5.7   Loops

Let's transition to `for` loops, which are a close relative of user-defined functions. Indeed, these are often used together, and can even be used to do similar things, with a few tweaks. We will see this in a moment. But let's start at the beginning. `for` loops allow for iterating some calculation or function over a bunch of different observations. So instead of typing out the same calculation line by line, while updating the main quantity of interest, you can just tell a `for` loop to do it for you (*pun not intended, but not regretted*). The syntax for `for` loops is similar to functions, where they begin with "for" and then start with some value in a sequence in parentheses. Then, within the braces, there is similarly a statement to be evaluated. Let's see how this works in practice by revisiting our temperature example.

```
fahrenheit <- c(60, 65, 70, 75, 80, 85, 90, 95, 100)

for (i in 1:length(fahrenheit)) {
  print((((fahrenheit[i] - 32) * 5) / 9)
}
```

```
## [1] 15.55556
## [1] 18.33333
## [1] 21.11111
```

```
## [1] 23.88889
## [1] 26.66667
## [1] 29.44444
## [1] 32.22222
## [1] 35
## [1] 37.77778
```

The same logic applies here, where we tell the loop to start at the first value (1) for each observation, i, in the vector of values in the object `fahrenheit`. And calculate the temperature conversion for each value in the `fahrenheit` vector. Finally, print the results for each supplied value. In sum, `for` loops are quite powerful tools that will significantly streamline your programming and make you think more efficiently in the process (e.g., "Rather than calculating values incrementally, how could I automate the process based on foundational logic/rules?").

With your knowledge of `if else` and `for` statements, you now know the foundational blocks of programming in R (or, really, any programming language). There are other programming structures, but conditional logic (`if else`) and looping (`for`) are the foundational components of programming.

5.7.1   Using a Loop to Test the Power of an Experiment

A common task for those conducting social science experiments is the calculation of the "power" of an experiment. In any experiment, we need to have enough participants to make sure we can detect a statistically significant effect (if any), but we do not want to make our sample size unnecessarily large, since this would waste time and money. Power tests are also a standard part of pre-registration, an increasingly common part of social science experiments.

One way to calculate the power of an experiment is to simulate what our data will look like. This involves drawing a sample from a particular (usually normal) distribution, creating a simulated treatment and control group, and running a statistical test on it. The proportion of the time that the test detects a statistically significant difference is the power of the test.

The code block below shows an adaptation of an example developed by the Evidence in Governance and Politics (EGAP) for those conducting experiments. It involves drawing a sample of 500 people from a normal distribution with a mean of 0 and a standard deviation of 1.[3] It then assigns 250 of the people to a treatment group that is, on average, 0.2 higher (i.e., the treatment effect is 2/10 of a standard deviation). It uses a t-test to test the statistical significance and records whether it was statistically significant at the 0.05 level ($p < 0.05$).

---

[3]Remember from your introduction to statistics that any normal distribution can be standardized to have a mean of 0 and a standard deviation of 1 by subtracting the mean and dividing by the standard deviation.

Note, in the chunk below, if assigned to the treatment group (`assignment ==` `1`), the outcome is `Y1`. If the observation is assigned to the control group (`1 -` `assignment == 1`), the outcome is `Y0`.

```r
N <- 500 # Number of participants in the study
alpha <- 0.05 # Level of significance set to p < 0.05
simulations <- 100 # The number of simulations we want
treatment_effect <- 0.2 # Expected effect of the experiment

experiment_result <- c() # Create empty vector for results

# Loop 100 times
for (i in 1:simulations) {
  Y0 <- rnorm(n = N, mean = 0, sd = 1) # random vals from normal
  Y1 <- Y0 + treatment_effect # exp outcome of treat
  assignment <- rbinom(n = N, size = 1, prob = .5) # treat/cont
  outcomes <- (Y0 * (1 - assignment)) + (Y1 * assignment)
  pvalue <- t.test(outcomes ~ assignment)$p.value # t-test for p
  significant <- ifelse(pvalue <= alpha, 1, 0)
  experiment_result <- c(experiment_result, significant)
}

mean(experiment_result) # Print the power of the experiment
```

```
## [1] 0.7
```

As you can see, the power of the test is about 0.6. Usually, for an experiment, we want to have at least 0.8 power. This suggests that we should add more cases to our experiment.

5.7.2 Using Loops to Explore Distributions

Suppose we drew a random sample of 50 respondents' self-reported political ideology on a 7-point scale, where 1 was extremely liberal and 7 was extremely conservative. The mean of that sample was 3.32, suggesting the average person in this sample sees themselves as generally moderate, or in the middle of the distribution of political ideology. Here is the code setting this up:

```r
sample_ideology <- c(3, 1, 2, 4, 4, 6, 1, 3, 2, 6,
                     1, 7, 3, 1, 4, 3, 4, 4, 1, 6,
                     7, 5, 7, 1, 1, 3, 2, 4, 1, 7,
                     1, 2, 1, 4, 6, 3, 2, 3, 1, 4,
                     1, 6, 3, 4, 5, 4, 1, 7, 2, 2)
mean(sample_ideology)
```

```
## [1] 3.32
```

Now, suppose we wanted to simulate the reported sample ideology to see whether this random sample was reflective of the broader American population. However, we are not exactly sure how many times to do this to accurately reflect the population of interest. To get traction on this question, the central limit theorem and law of large numbers can help us out. To see the shape distribution of many samples (central limit theorem) and how the location of the distributions change (law of large numbers), we can use a series of `for` loops, and plot the different distributions to see when and where (and whether) the samples converge on the underlying population.

`for` loops allow for iterating some calculation or function over many different observations. This is a simulation. The syntax of `for` loops is similar to functions, where they begin with "for" and then start with some value in a sequence in parentheses. Then, within the braces, there is a statement to be evaluated.

For each chunk below, we start by creating an empty vector in which to store our simulated values. We then specify the loop, to sample with replacement, based on the initially-drawn sample (`sample_ideology`), and then take the mean. We then plot each simulation and compare side by side via another tidy-friendly package covered in the previous visualization chapter, `patchwork`. The final result is in Figure 5.1.

```
# N = 5
sm1 <- rep(NA, 5)

for (i in 1:5) {
  samp <- sample(sample_ideology, 30, replace = TRUE)
  sm1[i] <- mean(samp)
}

# N = 20
sm2 <- rep(NA, 20)

for (i in 1:20) {
  samp <- sample(sample_ideology, 30, replace = TRUE)
  sm2[i] <- mean(samp)
}

# N = 50
sm3 <- rep(NA, 50)

for (i in 1:50) {
  samp <- sample(sample_ideology, 30, replace = TRUE)
  sm3[i] <- mean(samp)
}
```

```
# N = 100
sm4 <- rep(NA, 100)

for (i in 1:100) {
  samp <- sample(sample_ideology, 30, replace = TRUE)
  sm4[i] <- mean(samp)
}

# N = 500
sm5 <- rep(NA, 500)

for (i in 1:500) {
  samp <- sample(sample_ideology, 30, replace = TRUE)
  sm5[i] <- mean(samp)
}

# N = 1500
sm6 <- rep(NA, 1500)

for (i in 1:1500) {
  samp <- sample(sample_ideology, 30, replace = TRUE)
  sm6[i] <- mean(samp)
}

# N = 3500
sm7 <- rep(NA, 3500)

for (i in 1:3500) {
  samp <- sample(sample_ideology, 30, replace = TRUE)
  sm7[i] <- mean(samp)
}

# N = 7000
sm8 <- rep(NA, 7000)

for (i in 1:7000) {
  samp <- sample(sample_ideology, 30, replace = TRUE)
  sm8[i] <- mean(samp)
}

# Now plot each simulation
library(ggplot2)
library(patchwork)
```

```
p1 <- quickplot(sm1, geom="histogram", main="N=5", bins=30) +
  theme_minimal() +
  geom_vline(xintercept=3.32, linetype="dashed", color="red")

p2 <- quickplot(sm2, geom="histogram", main="N=20", bins=30) +
  theme_minimal() +
  geom_vline(xintercept=3.32, linetype="dashed", color="red")

p3 <- quickplot(sm3, geom="histogram", main="N=50", bins=30) +
  theme_minimal() +
  geom_vline(xintercept=3.32, linetype="dashed", color="red")

p4 <- quickplot(sm4, geom="histogram", main="N=100", bins=30) +
  theme_minimal() +
  geom_vline(xintercept=3.32, linetype="dashed", color="red")

p5 <- quickplot(sm5, geom="histogram", main="N=500", bins=30) +
  theme_minimal() +
  geom_vline(xintercept=3.32, linetype="dashed", color="red")

p6 <- quickplot(sm6, geom="histogram", main="N=1500", bins=30) +
  theme_minimal() +
  geom_vline(xintercept=3.32, linetype="dashed", color="red")

p7 <- quickplot(sm7, geom="histogram", main="N=3500", bins=30) +
  theme_minimal() +
  geom_vline(xintercept=3.32, linetype="dashed", color="red")

p8 <- quickplot(sm8, geom="histogram", main="N=7000", bins=30) +
  theme_minimal() +
  geom_vline(xintercept=3.32, linetype="dashed", color="red")

# piece ggplot objects together with the patchwork package
p1 +
  p2 +
  p3 +
  p4 +
  p5 +
  p6 +
  p7 +
  p8
```

Note that, as expected by the central limit theorem and the law of large numbers, as the sample size grows larger, the shape becomes more nor-

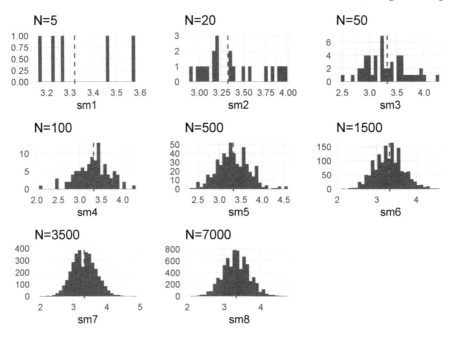

FIGURE 5.1
Simulation Results

mally/Gaussian distributed (via, central limit theorem) and the location of the distribution centers over the sample mean ideology of 3.32 (via, law of large numbers).

Indeed, this simple simulation shows that if we drew a large enough sample based on the original small sample, the shape and location of the distribution would indeed center over the "true population" value, suggesting our small initial sample was reflective of the true population. This is the idea behind a statistical technique generically referred to as "randomization distributions for statistical inference," which helps quantify evidence against some null hypothesis. Though the scope of this technique is beyond what we are interested in in this book, it remains a useful demonstration of a `for` loop for simulating these different sample sizes, but based on the original parameter values, which is a very common social science task.

Exercises

5.7.2.0.1 Easy

- Write a function to calculate body mass index (BMI) and store it in object, `bmi`. *Note*: the formula is $bmi = \frac{wt}{h^2}$, where wt is a person's weight (kilograms), and h^2 is a person's height squared (meters).

- Repeat the previous exercise, but use a `for` loop instead.
- Pass the following vectors, `weight` and `height`, to your function (from #1) and your loop (#2): `weight <- c(70, 75, 80, 60, 90)` and `height <- c(1.3, 2, 2.1, 1, 1.7)`. Do you see the same values returned? Why or why not?

5.7.2.0.2 *Intermediate*

- What are the two major types of operators and how do they differ?

5.7.2.0.3 *Advanced*

- Experiment with the experimental power loop above. How many cases should you have in order to get at least 80% power? What about if the treatment effect is 0.4?
- Try to nest the above loop inside another loop that tests the power with sample sizes of 100, 300, 500, 700, and 900. (*Hint*: you will have it loop over a vector `c(100, 300, 500, 700, 900)`).

5.7.3 Nesting Loops and Extreme Bounds Analysis (EBA)

As with conditional logic, we can nest loops to iterate over multiple vectors of values, or conduct several calculations. For example, the code block below shows a simple example of one loop nested in another to get all of the possible combinations of letters from two vectors.

```
vector1 <- c("a", "b", "c", "d")
vector2 <- c("a", "b", "c", "d")

for (i in 1:length(vector1)) {
  for (j in 1:length(vector2)) {
    print(paste("Value of vector1 is ", vector1[i],
                " and of vector2 is ", vector2[j])
          )
  }
}
```

```
## [1] "Value of vector1 is  a  and of vector2 is  a"
## [1] "Value of vector1 is  a  and of vector2 is  b"
## [1] "Value of vector1 is  a  and of vector2 is  c"
## [1] "Value of vector1 is  a  and of vector2 is  d"
## [1] "Value of vector1 is  b  and of vector2 is  a"
## [1] "Value of vector1 is  b  and of vector2 is  b"
## [1] "Value of vector1 is  b  and of vector2 is  c"
## [1] "Value of vector1 is  b  and of vector2 is  d"
## [1] "Value of vector1 is  c  and of vector2 is  a"
```

```
## [1] "Value of vector1 is  c  and of vector2 is  b"
## [1] "Value of vector1 is  c  and of vector2 is  c"
## [1] "Value of vector1 is  c  and of vector2 is  d"
## [1] "Value of vector1 is  d  and of vector2 is  a"
## [1] "Value of vector1 is  d  and of vector2 is  b"
## [1] "Value of vector1 is  d  and of vector2 is  c"
## [1] "Value of vector1 is  d  and of vector2 is  d"
```

This very simple nesting of loops to get all the combinations of items in several sets can be quite useful. To give just one example in applied research, it can be used in programming a technique called Extreme Bounds Analysis (EBA). EBA was a proposed technique for dealing with the issue of uncertainty in the specification of statistical models (Leamer, 1983, Leamer (2010)). The argument was that most scholarly papers only report a small subset of the statistical models they run, and that results (due to collinearity, missing data, or a number of other issues) may be different, depending on the variables included in the model. The proposal was for scholars to report the range of outcomes from every combination of three variables.

EBA has been used by a number of studies (Levine and Renelt, 1992, Xavier et al. (1997), Kennedy and Tiede (2013)), but has also been criticized by scholars who view it as atheoretical (McAleer et al., 1985). For our purposes, we care less about whether EBA is a desirable approach – it gives us a good example of nested loops.[4]

The chunks of code below show the code for a very simple EBA analysis of the ANES data for support of then-candidate Trump. First, we load the data – setting our working directory, loading the needed libraries, loading the data, and manipulating it into the form we want. We are looking at nine different features we think might affect support of Trump, meaning there are 84 different combinations of three features possible.

```
# Load needed libraries
library(tidyverse)
library(here)

# Set your working directory
setwd(choose.dir())

# Load the data via read_csv() and here()
NESdta <- read_csv(here("data", "anes_pilot_2016.csv"))

## Parsed with column specification:
## cols(
##    .default = col_double(),
```

---

[4]There is actually an R package now for EBA called Extremebounds that became available in 2016 (Hlavac, 2016).

```
##    version = col_character(),
##    pid2d = col_character(),
##    pid2r = col_character(),
##    other10_open = col_character(),
##    race_other = col_character(),
##    employ_t = col_character(),
##    religpew_t = col_character(),
##    disc_fed_disc_police_rnd = col_character(),
##    white_sections_rnd = col_character(),
##    lazy_violent_rnd = col_character(),
##    FEELING_THERMOMETER_rnd = col_character(),
##    meet_rnd = col_character(),
##    givefut_rnd = col_character(),
##    info_rnd = col_character(),
##    ISSUES_OC14_rnd = col_character(),
##    disc_selfsex_rnd = col_character(),
##    lazy_col_rnd = col_character(),
##    lazy_row_rnd = col_character(),
##    violent_col_rnd = col_character(),
##    violent_row_rnd = col_character()
##    # ... with 9 more columns
## )
## See spec(...) for full column specifications.
# Data manipulation for a simple analysis
NESdta_small <- NESdta %>%
  mutate(fttrump = ifelse(fttrump > 100, NA, fttrump),
         age = 2016 - birthyr,
         white = ifelse(race == 1, 1, 0),
         faminc = ifelse(faminc > 90, NA, faminc),
         republican = ifelse(pid3 == 2, 1, 0),
         religiosity = ifelse(pew_churatd>6, NA, 7-pew_churatd),
         news_interest = ifelse(newsint > 6, NA, 5 - newsint),
         conservativism = ifelse(ideo5 > 5, NA, ideo5),
         female = ifelse(gender == 2, 1, 0)) %>%
  dplyr::select(fttrump, age, white, faminc, republican,
                religiosity, news_interest, conservativism,
                educ, female) %>%
  dplyr::filter(!is.na(fttrump))


NESdta_small

## # A tibble: 1,197 x 10
##    fttrump   age white faminc republican religiosity
##      <dbl> <dbl> <dbl>  <dbl>      <dbl>       <dbl>
```

##	1	1	56	1	4	0	1
##	2	28	59	1	8	0	3
##	3	100	53	1	1	1	1
##	4	0	36	1	12	0	1
##	5	13	42	1	10	0	5
##	6	61	58	1	7	0	5
##	7	5	38	1	NA	0	5
##	8	85	65	1	10	1	1
##	9	70	43	1	8	0	5
##	10	5	80	1	10	0	6

```
## # ... with 1,187 more rows, and 4 more variables:
## #   news_interest <dbl>, conservativism <dbl>, educ <dbl>,
## #   female <dbl>
```

In the next chunk, we create our EBA function, eba(). The function takes as its input a data set, where the first column contains our dependent variable and the rest of the columns are the independent variables. There are three nested loops. The first (i), goes through the independent variables, starting with the second column of the data and moving to the third-from-last column. The second (j) goes from the third column to the next-to-last column. The third (k) goes from the fourth column to the last column. By doing this, we get all of the combinations of features. The first time through, it will get independent variables 1, 2, and 3. The second, it will get variables 1, 2, and 4. It will go through all the options for k, then will move to the second item in j (1, 3, 4; 1, 3, 5; etc.). For each of these, you can see that it runs a regression model for the three selected variables and saves the coefficients. The results are stored in a tibble, fullresults, and are returned by the function.

```
# Create a function that checks all combinations of 3 variables
eba <- function(dataset) {
  tempdata <- as.matrix(dataset)
  fullresults <- c()
  for (i in 2:(ncol(tempdata) - 2)) {
    for (j in (i + 1):(ncol(tempdata) - 1)) {
      for (k in (j + 1):ncol(tempdata)) {
        coefficients <- c(rep(NA, ncol(tempdata) - 1))
        tempModel <- lm(tempdata[,1] ~ tempdata[,i] +
                          tempdata[,j] + tempdata[,k]
                        )
        coefficients[(i - 1)] <- tempModel$coefficients[2]
        coefficients[(j - 1)] <- tempModel$coefficients[3]
        coefficients[(k - 1)] <- tempModel$coefficients[4]
        fullresults <- rbind(fullresults, coefficients)
      }
    }
  }
```

```
   fullresults <- as_tibble(data.frame(fullresults))
   names(fullresults) <- names(dataset)[2:length(names(dataset))]
   return(fullresults)
}
```

We can now call our `eba()` function using the ANES data we loaded previously. We pass the data set to the function and it returns the results. We can see the distribution for the variables `white` and `faminc` (family income) as examples.

```
trumpEBA <- eba(NESdta_small)
trumpEBA
```

```
## # A tibble: 84 x 9
##      age white  faminc republican religiosity news_interest
##    <dbl> <dbl>   <dbl>      <dbl>       <dbl>         <dbl>
## 1  0.341 14.3  -0.0619         NA          NA            NA
## 2  0.259  8.43 NA            32.4          NA            NA
## 3  0.303 14.5  NA              NA        3.32            NA
## 4  0.362 13.0  NA              NA          NA         -2.18
## 5  0.195  8.62 NA              NA          NA            NA
## 6  0.305 13.7  NA              NA          NA            NA
## 7  0.320 13.1  NA              NA          NA            NA
## 8  0.321 NA    -0.271        35.5          NA            NA
## 9  0.395 NA    -0.0607         NA        2.89            NA
## 10 0.443 NA     0.247          NA          NA         -2.36
## # ... with 74 more rows, and 3 more variables:
## #   conservativism <dbl>, educ <dbl>, female <dbl>
```

```
summary(trumpEBA$white)
```

```
##    Min. 1st Qu.  Median    Mean 3rd Qu.    Max.  NA's
##   7.134   9.799  13.078  12.582  15.551  17.898    56
```

```
summary(trumpEBA$faminc)
```

```
##      Min.  1st Qu.   Median     Mean  3rd Qu.     Max.
## -0.35179 -0.11555  0.01588  0.02031  0.15296  0.41478
##      NA's
##        56
```

Success! The `trumpEBA` tibble includes 84 observations, which is the number of feature combinations we expected. The results show clearly that white respondents report higher favorability of Trump, regardless of the other variables included as controls. Family income, however, has a positive effect on Trump's approval, on average, but appears more sensitive to model specification, with some specifications suggesting the opposite relationship.

Again, the point here is not about EBA as an approach, nor about the specifics of what increases support for then-candidate Trump. Rather, we are interested in showing how these nested loops can direct the computer to execute tasks quickly, which would otherwise take a long time to perform manually.

5.8   Mapping with `purrr`

The final concept we cover bridges programming and modeling. We are interested here in covering the `map()` family of functions, which is essentially a blend of loops and user-defined functions. Mapping functions offer users the ability to *map* or iteratively pass functions to values stored in arrays or vectors. For those familiar with the base R `apply` family of functions, mapping functions are essentially Tidyverse-flavored updates. The `map` family is housed in the `purrr` package for functional programming and is loaded when the `tidyverse` library is loaded.

We will use the smaller subset of the ANES data set created for the EBA example above. This time, we will include only the `female`, `fttrump`, and `birthyr` variables. We will then use a some tools learned in the *Data Management and Manipulation* chapter to `mutate()` `fttrump` (turning strange values to NAs) and `female` (recoding the levels of `gender`), and conclude by filtering out NAs, in line with the best practice of keeping data tidy.

```
NESdta_small <- NESdta %>%
  dplyr::select(gender, fttrump, birthyr) %>%
  mutate(fttrump = replace(fttrump, fttrump > 100, NA),
         female = ifelse(gender == 2, 1, 0)) %>%
  dplyr::filter(!is.na(fttrump))

NESdta_small <- NESdta_small %>%
  dplyr::select(-gender)

# inspect a small random sample to make sure things look good
sample_n(NESdta_small, 5)
```

```
## # A tibble: 5 x 3
##    fttrump birthyr female
##      <dbl>   <dbl>  <dbl>
## 1       68    1957      1
## 2      100    1982      1
## 3       91    1951      1
## 4        0    1940      1
## 5       22    1957      1
```

With our small data set built, we now use `split()` from base R to split our subsample into two groups: *female* (1) and *not female* (0). We store this in the new object `fems`. This exercise will demonstrate the simplest use of `map()`. The general syntax is to map a selected function (second argument) to the data object (first argument). So, here, we will pass the `nrow()` base R function and `summary()` base R function to the `fems` object. The output will be the number of observations (rows) associated with male and female respondents, and then a feature-level summary of observations in each level of `fems`, respectively.

```
# first split into groups using split from base R
fems <- split(NESdta_small, NESdta_small$female)

# Explore respondents in each group
map(fems, nrow)
```

```
## $`0`
## [1] 570
##
## $`1`
## [1] 627
```

```
# Now, feature level summary of each group
map(fems, summary)
```

```
## $`0`
##      fttrump          birthyr          female
##  Min.   :  0.0   Min.   :1921   Min.   :0
##  1st Qu.:  3.0   1st Qu.:1954   1st Qu.:0
##  Median : 39.0   Median :1970   Median :0
##  Mean   : 41.1   Mean   :1968   Mean   :0
##  3rd Qu.: 75.0   3rd Qu.:1982   3rd Qu.:0
##  Max.   :100.0   Max.   :1997   Max.   :0
##
## $`1`
##      fttrump           birthyr          female
##  Min.   :  0.00   Min.   :1924   Min.   :1
##  1st Qu.:  2.00   1st Qu.:1955   1st Qu.:1
##  Median : 20.00   Median :1965   Median :1
##  Mean   : 35.91   Mean   :1967   Mean   :1
##  3rd Qu.: 70.00   3rd Qu.:1982   3rd Qu.:1
##  Max.   :100.00   Max.   :1997   Max.   :1
```

Importantly, the basic `map` function previously used will always return a list, which is a *mixed* data type. But importantly, there may be many cases in which you would prefer working with a specific type of data or want a specific data type to condition the `map` function. If such is the case, then there are several other map functions, `map` that can be used. For example, the raw count of rows is an integer, so we could use `map_int()` to get the same result, but ensuring it's a real-valued integer.

```
map_int(fems, nrow) # note the different map() function
```

```
##   0   1
## 570 627
```

```
# make sure the returned value is an integer, NOT a list
is.integer(map_int(fems, nrow)) # return "TRUE"
```

```
## [1] TRUE
```

```
is.integer(map(fems, nrow)) # return "FALSE"
```

```
## [1] FALSE
```

With the logic under your belt, consider a couple extensions using `map`. First, rather than using `split()` from base R, the Tidyverse version of this is to "nest" via `nest()` or `unnest()`. Suppose we want to `nest()` respondents by "female or not" to return a *data frame* into a so-called "list-column", which can host multiple data types in a single vector.[5] In other words, we can nest a data frame *within* a row or column. List-columns from model output is a very common occurrence.

First, we will create a new object, `new_nes`, with data frames and/or tibbles for each level of `female`. The specific syntax below reads, "give me two rows and a data frame for the other two features in the small data set, nested by each level of `female`."

Then, we complicate matters for more descriptive output by adding two new features via `mutate()`: first, the number of rows/respondents, and second the mean values for Trump support from each group of respondents.[6]

```
(new_nes <- nest(NESdta_small, -female))
```

```
## Warning: All elements of `...` must be named.
## Did you want `data = c(fttrump, birthyr)`?
```

```
## # A tibble: 2 x 2
##   female data
##    <dbl> <list>
## 1      0 <tibble [570 x 2]>
## 2      1 <tibble [627 x 2]>
```

---

[5]The terminology is thanks to Jenny Bryan. See the `purrr` package documentation for more.

[6]Readers should note that the use of "doubles" above in `map_dbl()` indicate numeric data types for precision calculations, e.g., decimal places. They are used in many domains and for many tasks. See the `purrr` help documentation for more.

```
new_nes %>%
  mutate(n_row = map_int(data, nrow),
         mean = map_dbl(data, ~ mean(.x$fttrump)))
```

```
## # A tibble: 2 x 4
##   female data               n_row  mean
##    <dbl> <list>             <int> <dbl>
## 1      0 <tibble [570 x 2]>   570  41.1
## 2      1 <tibble [627 x 2]>   627  35.9
```

Now, reverse the nesting using **unnest()**, and we are back to where we started.

```
unnest(new_nes, data) %>%
  head()
```

```
## # A tibble: 6 x 3
##   female fttrump birthyr
##    <dbl>   <dbl>   <dbl>
## 1      0       1    1960
## 2      0     100    1963
## 3      0       0    1980
## 4      0      13    1974
## 5      0      61    1958
## 6      0       5    1978
```

5.9   Concluding Remarks

In this chapter, we learned some of the core building blocks of programming in both base R and the Tidyverse. All of these techniques and concepts are borrowed from other programming languages and adapted for the R language. For example, there are `for` loops (and also `repeat` and `while` loops) in Python, and different types of operators (e.g., logical) are used in virtually all programming languages like C and C#. With these tools in your toolbox, you can become a better, more efficient programmer, which will help you do a variety of tasks, whether writing R packages or conducting your own research.

6

Exploratory Data Analysis

The first thing any researcher should do prior to fitting models is get to know the data. This is the case because often the shape and structure of the data are unknown to the researcher. For example, if the data are skewed in a certain direction along some variable of interest, then this could limit the quality of inferences drawn after fitting a model (we will discuss this more below). But beyond overtly harmful effects, it is a good idea to know some basic features of the data as well as distributional shapes and patterns.

Broadly, this process of getting to know your data is called exploratory data analysis (EDA), and has it's roots in the work of John Tukey (Jones, 1987), who brought us many modern statistical tools for exploring data, such as the boxplot. Even though users are not fitting models in the predictive or causal ways of approaching analysis, exploring data is still very much *analysis* in that, by inspecting distributions and shapes of data, users are able to conceptualize trends and even generate baseline expectations, which will influence the research program in numerous ways downstream. Importantly, in other fields, such as machine learning, exploratory data analysis is closely linked with clustering, classification, and other useful techniques to help make sense of data when little of the data is known a priori. For example, machine learning researchers may fit a variety of clustering algorithms such as k-means, k-medoids, or CLARA (for big data applications) to pull out patterns and more precisely define groupings present within the data, but in a largely atheoretic way. Yet, in the social sciences, such approaches to exploring data are often avoided in an effort to guard against the possibility of ethical issues like searching for patterns too early, which could lead to a perception of "p-hacking" (searching for significant results based only on p-values across many model iterations) or post-hoc theorizing (suggesting you, the researcher, were aware of and anticipating the emergent patterns the whole time). Though there is a some degree of gray area in this regard, there could be a reasonable case made for a technique like clustering being helpful, not harmful, as a crucial step to contributing to a greater understanding of the non-random structure that is assumed to exist in data.

Our goal in this chapter, then, is to walk readers through a typical EDA project using the tools covered so far, while also bringing in a few new tidy techniques. This chapter will provide a roadmap for exploring data in search of descriptive trends, as well as offering guidance on how to discuss these trends in the context of a broader research project. Thus, while some new techniques and tools will be covered, the broader thrust of this chapter is in line with the tone of our book, where we are not interested in merely compiling a bunch of tidy functions, but rather are interested in demonstrating how to leverage the power of the tidyverse in the context of social science research. And importantly, every social science research project should include *some* EDA component.

In service of this goal, we will cover some common methods for exploring data including plots (bar plots, boxplots, and scatterplots), summary statistics (inter-quartile range (IQR), mean, median, minima and maxima, and so on), and a combination of these methods in tidy framework, drawing heavily from the `skimr` package.

Let's start by loading some useful packages and the data, followed by some quick tidying based on some techniques explored in the data management chapter.

```
# Load the libraries needed for this chapter
library(tidyverse)
library(here)
library(skimr)
library(amerika)

# Set the working directory
setwd(choose.dir())

# Load the ANES data and tidy a bit
NESdta <- read_csv(here("data", "anes_pilot_2016.csv"))

NESdta_sub <- NESdta %>%
  dplyr::select(fttrump, pid3, birthyr, gender, ftobama) %>%
  mutate(fttrump = replace(fttrump, fttrump > 100, NA),
         ftobama = replace(ftobama, ftobama == 998, NA),
         Party = case_when(pid3 == 1 ~ "Democrat",
                           pid3 == 2 ~ "Republican",
                           pid3 == 3 ~ "Independent")) %>%
  as.data.frame() %>%
  drop_na()
```

6.1   Visual Exploration

With the data loaded and tidied, let's start with basic visual descriptions of some variables of interest, which is the most common starting place in an EDA project given the rich descriptive nature of visualizations. A widely used visual tool for EDA is a bar plot, which is a close relative of the histogram, showing the categorical density of some variable of interest. These are especially useful in survey data where groupings of respondents are visually distributed across some variable of interest. Following the bar plot, we present Tukey's boxplot, which is more descriptive.

For these first two visual tools, building on the earlier *Visualization* chapter, we will rely on ggplot2 from the Tidyverse, and alter the geometric layers, geom_*, for bar plots first (geom_bar) and boxplots second (geom_boxplot). The result is in Figure 6.1.

```
ggplot(NESdta_sub, aes(fttrump, fill = Party)) +
  geom_bar(fill = amerika_palette(n = 233,
                                  name = "Dem_Ind_Rep7",
                                  type = "continuous")) +
  labs(x = "Trump Feeling Thermometer",
       y = "Count of Respondents",
       title = "Feeling Thermometer for Trump by Party",
       subtitle = "2016 ANES Pilot Study") +
  facet_wrap(~ Party) +
  theme_minimal()
```

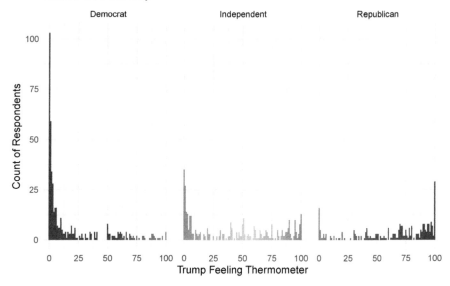

FIGURE 6.1
Feelings toward Trump by Party

In Figure 6.1, note a few expansions from the earlier visualization chapter. First, we are now using the `amerika` package, which is an American-politics inspired color palette generator for applications such as these (Waggoner, 2019). The idea is that applications map basic political knowledge to appropriate color palettes, where, for example, "Democrats" (or "liberals") are assumed to be on the left and thus blue, with "Independents" (or "moderates") in the middle and thus gray, followed by "Republicans" (or "conservatives") on the right, and thus red. Further, we use the powerful layer, `facet_wrap`, which allows for breaking up feelings toward Trump by party affiliation, based on the `pid3` variable from the ANES data. Each party is placed in a unique window, or "facet". This is a very useful tool for making an already descriptive plot *more* descriptive. Taken together, note in this bar plot that emergent patterns are in line with basic expectations of the distribution of political preferences in American politics, with Democrats (in blue) having strongest negative (or "coldest") feelings toward Trump, with Independents and Republicans having progressively more positive (or "warmer") feelings toward him. Building on these patterns from the basic bar plot, we can dig more into the numeric summaries of these data, but still using a visual tool: the boxplot.

Here, we introduce John Tukey's boxplot using the same two variables, `Party` and `fttrump` (feelings toward Trump). Boxplots are highly descriptive summaries of data of any size, showing the IQR, from the 1st quartile to the 3rd, in the box, with the line in the box representing the median of the distribution. The "whiskers" on the bottom and top of the plot show the minimum and maximum, respectively, of the data distribution. The dots represent outliers.

To build our boxplot, we simply change the geom to be `geom_boxplot` instead of `geom_bar` in the previous case. Here again we use the `amerika` package to quickly fill in appropriate colors in the boxes corresponding to each of the three major American political parties. The increase in descriptive information provided by boxplots allow for more thorough exploration of the data, especially in tandem with other visual tools such as the bar plot. These distributions in feelings toward Trump by party are shown in Figure 6.2.

```
ggplot(NESdta_sub, aes(x = Party, y = fttrump)) +
  geom_boxplot(fill = amerika_palette(name = "Dem_Ind_Rep3")) +
  labs(x = "Political Party",
       y = "Trump Feeling Thermometer Score",
       title = "Feeling Thermometer for Trump by Party",
       subtitle = "2016 ANES Pilot Study") +
  theme_minimal()
```

The boxplot shows that there are a lot of Democratic outliers given the very low mean and median as we might expect. This suggests that while the majority of the distribution of Democratic respondents has negative feelings toward Trump as we might expect, interestingly there are a few Democrats who think

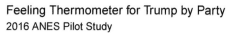

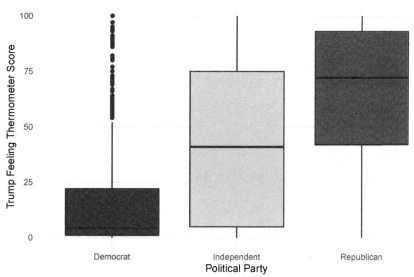

FIGURE 6.2
Feeling Distributions by Party

positively toward him. Further, in line with baseline expectations from our initial bar plot, we can see that the distribution of Republicans is situated at the highest end with the most positive feelings toward Trump, with Independents somewhere in between these major political parties. Regardless of the patterns, the point here is that these visual tools require relatively minimal code, but provide a great deal of information and nuance, which contribute to a greater understanding of our data.

We now transition to a third visual tool for exploring data, which is the scatterplot. The scatterplot is a slightly more intuitive approach to observing natural trends in the data. Similar to the previous two visual tools, `ggplot2` offers some excellent options for visualizing basic trends in data using a scatterplot. Here, as you might expect at this point, we simply need to update the 'geom_* layer to be `geom_point`. Building on the intuition of the bar plot, we can explore the range of feelings toward Trump (`fttrump`) by party affiliation (`Party`) in Figure 6.3, but also across the age of respondents (`birthyr`), allowing for even greater nuance to our descriptive exploration.

```
ggplot(NESdta_sub, aes(x = birthyr, y = fttrump,
                       color = factor(Party))) +
  geom_point() +
  scale_color_manual(name="Party",
```

```
                        values=amerika_palette(name="Dem_Ind_Rep3"))+
  labs(x = "Birth Year",
       y = "Trump Feeling Thermometer Score",
       title = "Feelings Toward Trump across Age and Party") +
  theme_minimal()
```

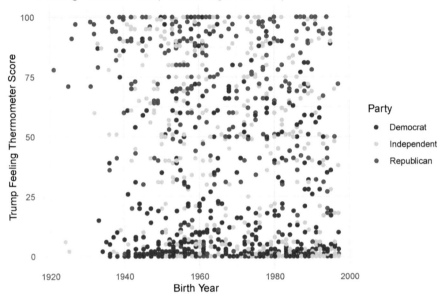

FIGURE 6.3
Feelings toward Trump by Party and Age

Conditioning the point colors by party affiliation, we can see several natural patterns emerge. First, note that many respondents across all parties had a relatively cold feelings toward Trump in 2016, seen in the tight grouping of points near 0.0 on the Y axis at the bottom of the plot. Inversely, there is also a high concentration of extremely positive feelings toward Trump, with points clustered around the top of the plot at 100 on the Y axis. Further, there are fewer respondents in between these extremes, suggesting Trump may be a polarizing political figure, where the majority of respondents either love him or hate him. Though more analysis would be needed to develop and test this idea, the point remains that an interesting natural pattern pointing to this possibility is present in these data, and was only uncovered by visually exploring our data.

While these trends may exist and while the conditional point color is useful, there are so many respondents that it is difficult to say much more about any possible trends based on this plot. To get a more precise, though still exploratory, look at these trends, we can add a nonparametric LOESS smoother, which simply describes the trend in the data. It is "nonparametric" in that it does not have an a-priori statistical definition of the data, such as that the relationship is linear. Rather, it simply describes conditional patterns based on *natural* values, thereby capturing *natural* variation.

To update our scatterplot to add these LOESS smoothers, we simply add another geometric layer, but this time it is called `geom_smooth`. Inside the layer, we must specify the method argument to be `loess`. Importantly, layering these smoothers after conditional colors already exist in the plot results in an automatic inheriting of the mapping aesthetic (see the earlier "Visualizing Your Data" chapter for more on mapping aesthetics). This means that the colors of the smoothers will *also* be conditional on party affiliation as we have specified, which will help us explore the data and trends consistently, and thus more efficiently. The result is in Figure 6.4.

```
ggplot(NESdta_sub, aes(x = birthyr, y = fttrump,
                       color = factor(Party))) +
  geom_point() +
  geom_smooth(method = "loess", se = FALSE) +
  scale_color_manual(name="Party",
                     values=amerika_palette(name="Dem_Ind_Rep3"))+
  labs(x = "Birth Year",
       y = "Trump Feeling Thermometer Score",
       title = "Feelings Toward Trump across Age and Party") +
  theme_minimal()
```

`geom_smooth()` using formula 'y ~ x'

In addition to corroborating earlier patterns of extremity and intuition across major American political parties, the addition of the LOESS smoother reveals another interesting pattern which is that across all parties, younger respondents (toward the right of the X axis along "Birth Year") all favor Trump much less than their older counterparts to the left of the X axis. All smoothers start at a higher point on the left of the X axis than where they end on the right of the X axis along `birthyr`, though some variation seems to spike in the middle-range of respondents.

As demonstrated with these few simple visual techniques, it is clear that the Tidyverse has many powerful visual tools available for exploring natural trends in data.

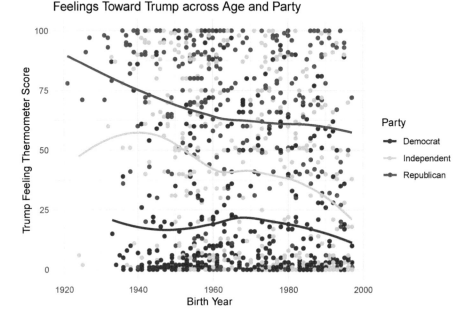

FIGURE 6.4
Smoothed Feelings toward Trump

Exercises

6.1.0.0.1   Easy

- Plot support for Trump (`fttrump`) as both a bar plot and a histogram by altering the `geom_*`. Place these plots side by side. What do you see? How do these plotting methods differ, and when might it be more or less appropriate than another?
- Do a quick Google search for the `RColorBrewer` package. How many palettes are included in this package? Update the plot above of `birthyr` by `fttrump`, colored by `Party` using the `Dark2` palette from `RColorBrewer`. Play around with other color palettes in both the `RColorBrewer` and `amerika` packages.

6.1.0.0.2   Intermediate

- What is a boxplot and how is it useful?
- Suppose your teacher requested a "full visual exploratory story." Where would you start, what techniques would you include, and how would you organize this "story told with data"?

6.1.0.0.3 Advanced

- What did Tukey mean by needing both "confirmatory" and "exploratory" hypotheses? Recall Tukey was writing these things in the 1970s and 1980s. How then, if at all, does his conclusion relate to modern predictive modeling and data science?
- Create an interactive version of the previous scatterplot of feelings toward Trump over the range of age by political party affiliation.

6.2 Numeric Exploration

While visual exploration is an important first step to take prior to fitting models, it is not the only exploratory tool at our disposal. Numeric summaries and descriptions of data are also quite useful for unpacking and exploring data efficiently. Base R has many useful tools for this, including summary(). We will start here, but bring in the tidy approach quickly, which allows for more efficient front-end filtering and wrangling as we saw in the earlier data management chapter. Consider first some basic summary statistics calculated using the summary() command in base R.

```
summary(NESdta_sub)
```

```
##      fttrump            pid3             birthyr
## Min.    :  0.0    Min.    :1.000    Min.    :1921
## 1st Qu.:  2.0    1st Qu.:1.000    1st Qu.:1954
## Median : 28.0    Median :2.000    Median :1967
## Mean    : 37.9    Mean    :1.931    Mean    :1968
## 3rd Qu.: 71.0    3rd Qu.:3.000    3rd Qu.:1982
## Max.    :100.0    Max.    :3.000    Max.    :1997
##      gender            ftobama            Party
## Min.    :1.000    Min.    :  0.00    Length:1115
## 1st Qu.:1.000    1st Qu.:  6.00    Class :character
## Median :2.000    Median : 55.00    Mode  :character
## Mean    :1.529    Mean    : 49.62
## 3rd Qu.:2.000    3rd Qu.: 88.00
## Max.    :2.000    Max.    :100.00
```

The output from this object includes variable-level numeric summaries consisting of: minimum value, 1st quartile, median, 3rd quartile, mean, and maximum (as well as a count of missing values (NAs), if those exist in the data). When calling the summary on the full data set, the output produced these basic summary statistics for *all* variables in the data set, which here was our tidied NESdta_sub data object. However, we urge caution in such a use of summary(), as some values may not make sense. For example, R will calculate the mean value of a categorical dummy variable, which takes on only values of 0 and 1. Even though a "mean" is calculated, this has no substantive meaning.

Though a useful starting place, the tidy approach to numeric exploration of data is much more efficient and cleanly output. Though there are many tools in the tidyverse that we could use, we will focus on: `sample_n`, `filter`, `group_by`, and `skim` (from the `skimr` package, which is written to complement the tidy approach, as we will see in a moment).

If we were interested in only grabbing a subset of rows/observations from the full data set, but wanted it to be a random grab to get a "fair" (or perhaps, *fairer*) look at the data, the `sample_n` function from the Tidyverse is a good place to start. It has a number of useful arguments, such as allowing the user to specify how many random observations to grab (`size`) as well as whether to sample with or without replacement (`replace`). Consider the following example, inspecting a random sample with replacement of observations of length 10 across all variables in our `NESdta_sub` data object.

```
sample_n(NESdta_sub,
         size = 10,
         replace = TRUE)
```

```
##      fttrump pid3 birthyr gender ftobama       Party
## 1         98    2    1931      1       0  Republican
## 2         84    2    1978      1       0  Republican
## 3         74    2    1962      2      53  Republican
## 4         85    2    1951      1       0  Republican
## 5         18    3    1996      1      61 Independent
## 6         12    3    1958      1       0 Independent
## 7         99    2    1940      1      10  Republican
## 8         58    1    1992      2      43    Democrat
## 9         17    3    1967      1      31 Independent
## 10        95    3    1946      1       1 Independent
```

We get a cleanly formatted tibble of 10 randomly selected observations/respondents across all variables in our `NESdta_sub` object.[1]

In line with the tidy approach to programming, we can layer several functions using the pipe operator (`%>%`) we previously discussed in the data management chapter, as well as at the outset of this chapter when tidying and creating our restricted data object, `NESdta_sub`. For example, we may want to explore a random set of observations that appear in the data after a specific date. In

---

[1] We encourage users to adjust and alter the arguments in the function to observe how the random grab change each time the function is called, e.g., changing `size` from 10 to 30, or setting `replace = TRUE`. Or at a minimum, consider testing the randomness claim here by simply running the previous code chunk again (and again) seeing how the sampled rows differ each time.

this case, we would pipe the `filter` function to restrict our small sample of length 5 to include respondents younger than the median birth year, which is 1967.[2] Building on the discussion of the `filter()` function in the earlier data management chapter, it is useful to point out that there are other versions of `filter()`, which allow for conditional filtering of data, or filtering based on specific values of a given variable, e.g., `filter_if()` or `filter_at()`. These can be extremely useful in numerically exploring specific chunks of the data, or data based on some condition of interest as in our example here. We encourage users to inspect the `dplyr` package documentation for many more details on the wide array of options available in the `filter` family, let alone the full range of munging functions in the `dplyr` package. To do so, run the command `?filter` with a single ? for the specific function, or `??dplyr` with two ?? to inspect documentation for the entire package.

```
median(NESdta_sub$birthyr) # 1967
```

```
## [1] 1967
```

```
NESdta_sub %>%
  filter(birthyr > median(birthyr, na.rm = TRUE)) %>%
  sample_n(5, replace = TRUE)
```

```
##   fttrump pid3 birthyr gender ftobama      Party
## 1      98    2    1980      1      25 Republican
## 2       2    1    1972      2     100   Democrat
## 3       3    1    1986      2     100   Democrat
## 4       1    2    1981      2       1 Republican
## 5       1    1    1983      1      94   Democrat
```

Similarly, we could group observations along a specific attribute by piping another layer using the `group_by()` function, and then drawing a random sample of 5 from each party, again, for all respondents younger than the median age in the sample.

```
NESdta_sub %>%
  filter(birthyr > median(birthyr, na.rm = TRUE)) %>%
  group_by(pid3) %>%
  sample_n(5, replace = TRUE)
```

```
## # A tibble: 15 x 6
## # Groups:   pid3 [3]
##   fttrump  pid3 birthyr gender ftobama Party
##     <dbl> <dbl>   <dbl>  <dbl>   <dbl> <chr>
## 1       1     1    1983      1     100 Democrat
```

---

[2] Note the `na.rm` argument, which in this case is set to `TRUE`. This simply means that we would like to filter values at the supplied threshold for all observations containing real values, not those with missing values.

```
##  2      40      1      1980      1      80 Democrat
##  3       3      1      1987      2      75 Democrat
##  4      39      1      1988      1      90 Democrat
##  5      50      1      1978      1      51 Democrat
##  6      80      2      1971      1       3 Republican
##  7       0      2      1972      1       4 Republican
##  8      80      2      1982      1       3 Republican
##  9       0      2      1987      2      91 Republican
## 10      91      2      1976      1      12 Republican
## 11       3      3      1973      2      97 Independent
## 12       3      3      1997      2      15 Independent
## 13       5      3      1979      1      15 Independent
## 14      33      3      1981      1      39 Independent
## 15      87      3      1968      2       3 Independent
```

Importantly, in all of these exercises of exploring the data as well as those discussed in the *Data Management and Manipulation* chapter, you can store these restricted data sets as objects, as with any value in R. Recall, as we noted in the Foundations chapter, that R is built around the notion of "object-oriented programming", where storing values in objects is at the heart of working in R. And recall that objects are created by simply passing one value to another through the assignment operator, <-.

As an aside, the intuition and consistency of the Tidyverse should hopefully be apparent by this point in the book. To reiterate, the aim of tidy programming is to make programming in R as simple, concise, clear, and consistent as possible. For example, in many of the Tidyverse packages, you will see a lot of similarities in the names of arguments and functions, e.g., _all and _by. These suffixes appear in many places and mean exactly what they imply: "apply this function *by* (based on) a given value" or "do this thing for *all* values in the variable or for *all* variables in the data set. The result is these tools are useful for both EDA as well as streamlining programming and workflows for more productive analysis in R.

Exercises

6.2.0.0.1   Easy

- Calculate the mean of `birthyr` and then the median of `birthyr`. How might our view of the data change when inspecting each of these values? And more importantly, what picture of the data are each of these numeric descriptors providing?
- What does the "mapping" argument do in any `ggplot()`? (*hint*: consider ?)

6.2.0.0.2 Intermediate

- Create a new variable called `dem_mean_birth` that records the mean `birthyr` for all Democrats in the data. Do the same two more times for Republicans and Independents, respectively, altering the variable name as it makes sense (e.g., `rep_mean_birth` for Republicans). Display these in a tibble and discuss substantive patterns and differences (if any) you see.
- What advantage does a boxplot offer over a scatterplot and how might this impact exploratory conclusions drawn?

6.2.0.0.3 Advanced

- Building on the discussion of the boxplot above, what *numeric* value is revealed in a boxplot? Do you think the **numeric** presentation of these values is more effective and descriptive than the **visual** presentation, or vice versa? Why?
- Write a function to take on a vector of feeling thermometer ratings and automatically generate a scatterplot over the range of age. Using this function, plot feelings toward all political candidates over the range of age. Placing these in a grid (*hint*: consider the `gridExtra` package for multiple `ggplot()` objects), what are some general patterns you see relating to feelings and respondents' ages, or are there any trends?

6.3 Putting it All Together: Skimming Data

Beyond addressing isolated powerful Tidyverse tools that can be used for exploratory data analysis prior to fitting models, we can combine these visual and numeric tidy functions for an even cleaner and simpler look at the data. To do so, we will rely on the `skim()` function from the `skimr` package.

The `skim()` function can be used for summary statistics for individual variables or entire data sets. Though more informative and useful than the `summary()` function in base R for a variety of reasons, one of the most powerful extensions of `skim()` is the separation of variables in a data set by variable type (e.g., factor, numeric, character, etc.). Upon distinguishing between variable type, `skim()` presents summary statistics by variable that make sense (e.g., bypassing the meaningless "mean" calculation for dummy variables mentioned above), in addition to a visual of the distribution of each variable in the data. Consider the following exercise of skimming the variables in our restricted `NESdta_sub` data object.

```
skim(NESdta_sub)
```

In addition to the many useful summary statistics by variable type as well as the histogram of the variable's distribution, the standard deviations for all numeric variables is included, but not for categorical or character variables, as this calculation would not make any sense. Further, the complete and missing values are quite useful in contexts where little is known about the data or when the data are particularly large and messy. Regarding different syntax, instead of `minimum`, `median`, and so on in the `summary()` function, `skim()` calls the quantiles `p0`, `p25`, `p50`, etc. The values remain the same, despite the different terminology.

Inspecting our data set, a few things stand out. First, we have no missing observations. Also, inspecting the histogram for `birthyr`, for example, we see that it is skewed toward the younger end, where we have far fewer older respondents than young respondents.

Though already significantly more informative, we can go farther in skimming our data given that, as previously mentioned, `skimr` was designed to fully integrate with tidy programming, seen, for example, in the reliance on tidy vocabulary.[3]

Exercises

6.3.0.0.1   Easy

- Use the `skim` function to numerically explore *all* feeling thermometers in the `NESdta` data set. (*hint*: think back to the Data Munging chapter on selecting subsets of variables that start with a common string, like, e.g., "ft" for feeling thermometer).
- What does it mean for a package like `skimr` to be "complementary of the Tidyverse"?

6.3.0.0.2   Intermediate

- Pick any three variables from the `NESdta` data set, and "tell a descriptive story" with these data. In other words, using the exploratory techniques discussed in this chapter, how would you visually and numerically explore and present these data to a general audience?
- Plot a random sample of the Obama feeling thermometer ratings (`ftobama`) of `size = 50`, conditional on gender. Overlay a loess smoother. What do you see? What does the loess smoother tell you?

---

[3]To further illustrate this point, users can even specify tidyverse commands with a `skim` call, e.g., `skim(ANES, starts_with("ft"))`, which would display the summary statistics and histograms for all feeling thermometers in the ANES data set (i.e., beginning with "ft" prefix).

6.3.0.0.3 Advanced

- Suppose you saw a pattern that surprised you, like, e.g., more Republican support for Obama than among Democratic respondents. How would you investigate this seemingly odd pattern?
- Suppose you plotted the distribution of feelings toward Trump (`fttrump`), and saw a big spike in support at the value of 998. What would this tell you and how would you know? What would be some exploratory follow-up steps you could take in response?

6.4 Concluding Remarks

In this chapter, we covered how to visually and numerically explore data in line with the Tidyverse approach to programming in R. This approach leverages consistent vocabulary across a variety of functions to result in cleaner code that is simpler to link, layer, and update.

Of note, we highly recommend readers explore the many options available in the `skimr` package, as well as combine functions and operations from other Tidyverse packages using the `%>%`. As noted throughout, piping functions that are built using the same vocabulary will minimize the steepness of the learning curve of working in R.

In the next and final substantive chapter we build on the principles of EDA covered here, and transition to statistically modeling and visualizing relationships in a tidy way.

7

Essential Statistical Modeling

When approaching statistical modeling in the social sciences, we most often operate from the "null hypothesis testing framework" (i.e., NHST), where we are interested in addressing the question, *Can we reject the "null hypothesis of no effect" given the data we observe, or not?*

In this chapter, we will address this question for several very common situations. We start with an example of one-sample and two-sample t-tests. Next, we continue with the exploration of cross-tabulation tables, showing how to find the chi-square value. Third, we explore the very versatile methods of correlation and ordinary least squares regression (OLS). Finally, for a binary response variable we demonstrate logistic and probit regression.

In this chapter we walk through essential techniques for fitting, interpreting, and diagnosing each of these commonly used modeling techniques from a Tidyverse perspective. Importantly, this is *not* a statistics text, but rather an introduction to R and the Tidyverse for social scientists. As statistical modeling is essential to social scientists, we cover these topics, but only at a high level and with a greater focus on fitting widely used models in the Tidyverse. The expectation is that the reader will have at least a passing familiarity with the statistical techniques discussed below. If the reader needs a review, there are plenty of excellent introductions to statistics for the social sciences (Finlay and Agresti, 1986, Gailmard (2014)).

Once you have the basics of model fitting in R down, you will find that these patterns tend to persist as you move to trying different models and techniques.

7.1 Loading and Inspecting the Data

As always, we start this section by starting RStudio and setting our working directory or opening the R project, .Rproj, file you will be using.

```
# Set your working directory
setwd(choose.dir())
```

We continue with our ANES data set in this chapter. In this chapter we will focus on a constrained set of variables:

1. `fttrump`: Feeling thermometer for Trump in 2016 (from 1 to 100, where 1 = cold and 100 = warm)
2. `pid3`: Respondent's party affiliation (1 = Democrat, 2 = Independent, 3 = Republican)
3. `birthyr`: Respondent's birth year
4. `gender`: Respondent's gender (1 = male and 2 = female)
5. `ftobama`: Feeling thermometer for Obama in 2016 (from 1 to 100, where 1 = cold and 100 = warm)

First, we need to load some relevant packages and load the corresponding libraries.

```
library(tidyverse)
library(here)
library(corrr)
library(skimr)
library(amerika)
library(broom)
library(rstatix)
library(janitor)
library(performance)
library(see)
```

A few of these packages are worth noting, since they are new to this chapter. The `corrr` package provides functions to evaluation correlations within a tidyverse framework. `rstatix` does the same for basic statistical functions and will be used for t-tests below. We will again be using the `janitor` package for analyzing cross-tabulation – in this case, analyzing chi-squared.

With the packages loaded, we now load our data, `NESdta`, using the `here` package. Next, we create a new data object, `NESdta_sub`, with only these 5 variables of interest, and do a bit of cleaning using the functions we learned in the data management chapter. Also, note that we are creating a new variable, `Party` by recoding the `pid3` variable to correspond with the actual party labels, instead of 1, 2, and 3. This will come in handy for plots below.

```
NESdta <- read_csv(here("data", "anes_pilot_2016.csv"))

NESdta_sub <- NESdta %>%
  dplyr::select(fttrump, pid3, birthyr, gender, ftobama) %>%
  mutate(fttrump = replace(fttrump, fttrump > 100, NA),
         ftobama = replace(ftobama, ftobama == 998, NA),
         Party = case_when(pid3 == 1 ~ "Democrat",
```

```
                            pid3 == 2 ~ "Republican",
                            pid3 == 3 ~ "Independent"),
            female = ifelse(gender == 2, 1, 0)) %>%
    as.data.frame() %>%
    drop_na()
```

Before moving into the actual tests, we will use the `glimpse()` function we used earlier to check if the data looks like what we expect.

```
glimpse(NESdta_sub)
```

```
## Rows: 1,115
## Columns: 7
## $ fttrump <dbl> 1, 28, 100, 0, 61, 5, 85, 70, 5, 74, 95...
## $ pid3    <dbl> 1, 3, 2, 1, 3, 1, 2, 3, 1, 2, 3, 1, 2, ...
## $ birthyr <dbl> 1960, 1957, 1963, 1980, 1958, 1978, 195...
## $ gender  <dbl> 1, 2, 1, 1, 1, 1, 1, 1, 1, 1, 2, 1, 2, ...
## $ ftobama <dbl> 100, 39, 1, 89, 0, 73, 0, 12, 87, 32, 1...
## $ Party   <chr> "Democrat", "Independent", "Republican"...
## $ female  <dbl> 0, 1, 0, 0, 0, 0, 0, 0, 0, 0, 1, 0, 1, ...
```

Now that the data is ready, let's start producing some statistical models.

7.2   t-statistics

One of the most basic statistical tests is a difference in means test. If we have two groups (for example, a treatment group and a control group), we can compare the means between the two (or more) groups using a t-distribution. These tests are often used in combination with the difference in means tables we demonstrated in the data management chapter. These t-tests are quite easy to calculate in R, as we demonstrate below.

There are several different types of t-tests. We start with the simplest, a one-sample comparison of means. In this situation, we are comparing the mean in a sample against a hypothetical population mean. So let's say we want to test whether approval for then-candidate Donald Trump is above 50 on a 100 point scale. We might interpret this as the point where people view him more positively than negatively.

The `t_test()` function from the `rstatix` package is used for this purpose. There are two baseline arguments that are required. The first is the function to be analyzed. The ~ operator is often used in statistical tests and is often read as "is approximated by." In this case, we are evaluating one group against a null model so we place the variable to be evaluated on the left-hand side, followed by ~ 1. The other part of this test is to specify the hypothetical population mean we are testing with `mu =`. In this case, we are testing the likelihood that the population mean for Trump's approval is 50, given the data from the ANES sample. Finally, we also specify `detailed = TRUE` to get a detailed report of the results that includes the 95% confidence intervals.

```
NESdta_sub %>%
  summarize(mean_approval = mean(fttrump, na.rm = T))
```

```
##    mean_approval
## 1      37.89596
```

```
# t-test fit
NESdta_sub %>%
  t_test(fttrump ~ 1, mu = 50, detailed = TRUE)
```

```
## # A tibble: 1 x 12
##    estimate .y.   group1 group2     n statistic          p
## *     <dbl> <chr> <chr>  <chr>  <int>     <dbl>      <dbl>
## 1      37.9 fttr~ 1      null ~  1115     -11.1 4.45e-27
## # ... with 5 more variables: df <dbl>, conf.low <dbl>,
## #   conf.high <dbl>, method <chr>, alternative <chr>
```

In the first chunk of code, we show that the average approval for Trump in this survey was well below 50 – in fact it was about 37.9 – using the `summarize()` function we learned earlier.

Upon conducting the t-test, we unsurprisingly found that the difference between the mean we find in the sample and the hypothetical mean of 50 is statistically significant. The t-statistic is -11.07 and the p-value is less than 0.001, both indicating that the difference in means is greater than that which is typically considered "statistically significant." We can also see this in the 95% confidence intervals that range from 35.7 to 40.0, far below the hypothetical mean of 50.

The default for this test is a two-tailed test, but this can be changed by specifying `alternative = "greater"` for a right-tailed test or `alternative = "less"` for a left-tailed test. We can also change the confidence level we desire by changing the `conf.level` option. For example, to do the test at the 99% level of confidence, we would specify `conf.level = 0.99`.

The next test is to compare two groups. So, for example, let's say we want to know if the difference in Trump's approval between men and women is statistically significant at this point in the 2016 campaign. We can use the same `t_test()` function. In this case, we will specify the groups we want to test. We will use the `female` variable, which is 1 if the respondent is female and 0 otherwise. We specify the group by changing ~ 1 to ~ `female`. The rest of the example remains the same.

```
NESdta_sub %>%
  group_by(female) %>%
  summarize(avg_approval = mean(fttrump, na.rm = TRUE))
```

```
## # A tibble: 2 x 2
##    female avg_approval
##     <dbl>        <dbl>
## 1       0         40.9
## 2       1         35.2
```

```
# t-test fit
NESdta_sub %>%
  t_test(fttrump ~ female, detailed = TRUE)
```

```
## # A tibble: 1 x 15
##    estimate estimate1 estimate2 .y.   group1 group2    n1
## *     <dbl>     <dbl>     <dbl> <chr> <chr>  <chr>  <int>
## 1      5.70      40.9      35.2 fttr~ 0      1        525
## # ... with 8 more variables: n2 <int>, statistic <dbl>,
## #    p <dbl>, df <dbl>, conf.low <dbl>, conf.high <dbl>,
## #    method <chr>, alternative <chr>
```

Again, we start by looking at the difference between the two group means using the `group_by()` and `summarize()` functions from earlier in the book. It appears that, on average, women give Trump about a 5 point lower rating than men.

The t-test shows that this difference is statistically significant. The 95% confidence interval of this difference suggests that there is between a 1 and 10 point difference between men and women in this sample. The p-value is 0.009, which is well below the usual 0.05 level of significance.

As with the one-sample t-test, the options will allow you to change the confidence levels or move to a one-tailed test. By default, the `t_test()` function for difference between groups assumes that the groups have different variances, but this can be changed by specifying `var.equal = TRUE`. Similarly, by default, the function assumes that the groups are not paired, but this can be changed by specifying `paired = TRUE`.

We have only scratched the surface of the options available for conducting t-tests in R, and the associated plot options and diagnostic tests. This, however, gives you the foundation to find out more on your own about how to conduct t-tests.

7.3   Chi-square Test for Contingency Tables

In the data management chapter, we also introduced how to create contingency tables in R. Now we want to know if the differences observed in the contingency tables are statistically significant, or if they might be due to random sampling error.

As a reminder, here is a simple way to get a cross-tab using the `janitor` package's `tabyl()` function. In this case, we create a cross-tabulation of the respondent's political party ID with their stated gender (where 1 indicates female and 0 indicates male).

```
NESdta_sub %>%
  tabyl(Party, female)
```

```
##          Party   0   1
##        Democrat 188 268
##     Independent 208 171
##      Republican 129 151
```

Notice that this does not lend itself to a comparison of means because both variables are nominal/categorical. We will instead use the chi-squared test of statistical significance, which is executed with the `chisq.test()` function.

```
NESdta_sub %>%
  tabyl(Party, female) %>%
  chisq.test()
```

```
##
##   Pearson's Chi-squared test
##
## data:    .
## X-squared = 15.64, df = 2, p-value = 0.0004017
```

The results show that the differences in party affiliations between men and women are much higher than we would have expected by random chance. The chi-squared value is 15.64, and the corresponding p-value is 0.0004 – much lower than the standard 0.05 level of confidence often used as a threshold in the social sciences.

7.4 Correlation

We now move to a discussion of correlation. Correlation provides information about the direction (positive or negative) and strength of a linear relationship between two variables.

Before calculating the correlation, however, it is always valuable to inspect your data as we discussed in the previous *Exploratory Data Analysis* chapter. Though there are many ways to do this, we will focus here on visualization as this offers a more intuitive, clean look at the distribution of our variables of interest.

To do so, we start with a scatterplot of the distribution of feelings toward Trump plotted against the distribution of feelings toward Obama.[1] As such, with this plot in Figure 7.1, we can get a first look at whether respondents naturally vary in preferences for candidates of different parties, as we might expect they would.

```
ggplot(NESdta_sub, aes(fttrump, ftobama)) +
  geom_point(alpha = 0.7, color = "Midnight Blue") +
  labs(x = "Trump Feeling Thermometer",
       y = "Obama Feeling Thermometer") +
  theme_minimal()
```

Sure enough, we can see clusters of respondents in the upper left and lower right corners of the plot, suggesting that respondents who really favor Obama (higher values on the Y axis) tend to also really oppose Trump (lower values on the X axis). While the same is true for the opposite in the lower right corner, its not as stark as we might expect. We will explicitly explore the role of partisanship in this story later in the chapter.

A natural next step to see how strong the relationship is between these two variables is to check the correlation between them. Correlation is also often used to diagnose collinearity and other issues (discussed more below) in regression models. Pearson's correlation coefficient, ρ, which is the most commonly used, ranges from -1 for a perfect negative correlation to 1 for a perfect positive correlation, with 0 indicating no correlation. The `corr` package provides a range of highly useful modifications to the standard R correlation function so we will be leveraging its `correlate()` function.

So let's select the variables plotted above and find their correlation. This can be accomplished with the following code.

---

[1]Note: the `alpha` argument in the `geom_point()` function sets the transparency of the points, where values $<$ 1 produce some amount of transparency, and values $=$ 1 produce fully filled in points.

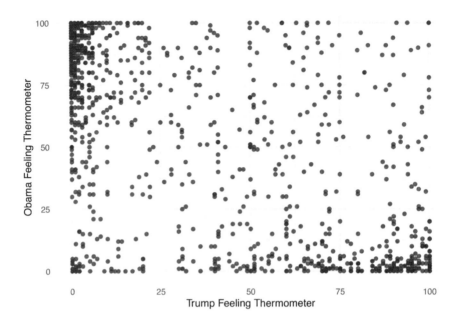

FIGURE 7.1
Feelings toward Trump and Obama

```
correlation <- NESdta_sub %>%
  dplyr::select(ftobama, fttrump, birthyr) %>%
  correlate()
```

```
##
## Correlation method: 'pearson'
## Missing treated using: 'pairwise.complete.obs'
```

```
correlation
```

```
## # A tibble: 3 x 4
##    rowname ftobama fttrump birthyr
##    <chr>     <dbl>   <dbl>   <dbl>
## 1 ftobama  NA      -0.593   0.149
## 2 fttrump  -0.593  NA      -0.165
## 3 birthyr   0.149  -0.165  NA
```

The resulting tibble shows that there is a moderate negative relationship
between approval of Trump and approval of Obama. It also shows a small
positive correlation between birth year and support of Obama (younger people
give him a higher rating), and a small negative correlation for approval of
Trump (older people give him a higher rating).

There is a lot more that can be done with the `corr` package, in part because of its tidyverse setup. For those interested in finding out more of what can be done with correlations, the author of the package has provided an excellent summary of his frustration with the base R correlation functions and the capabilities of the `corr` package here.

Exercises

7.4.0.0.1 Easy

- Calculate the correlation between a person's birth year, `birthyr`, and their approval of both Obama and Trump, `ftobama` and `fttrump`, respectively.
- Using the scatterplot code we developed in the "Exploratory Data Analysis" chapter, plot the relationship between support for Obama and Trump. Using `geom_smooth()`, check whether the relationship looks linear.

7.4.0.0.2 Intermediate

- How, if at all, is correlation related to regression?
- Sometimes two libraries will contain functions with the same name. This is the case with `skimr` and `dplyr`, which both have a `filter()` function. Try using the `filter()` function without specifying the library. How would you update your code to avoid this error?

7.4.0.0.3 Advanced

- Describe a scenario where it would *not* make sense to calculate a correlation coefficient.
- Suppose you get a correlation of 1.0. What would this tell you and what might be some follow-up steps you would take to investigate?

7.5 Ordinary Least Squares Regression

With our data loaded and explored, as well as a quick check for correlations between variables of interest, we can now fit a simple bivariate linear ordinary least squares regression (OLS) model, predicting feelings toward Trump (`fttrump`) as a function of respondents' ages (`birthyr`).

OLS is a very powerful and flexible model, that is used in a variety of circumstances. Like correlation, a basic OLS model assumes that there is a linear relationship between the independent and dependent variable. This assumption can, however, be relaxed by adding squared, cubic, or even higher order exponents to the regression equation. Basic OLS also assumes that the dependent variable is continuous, but this model is sometimes used with ordinal (or even dichotomous) data.

For simplicity of demonstration, we are going to assume a linear relationship between respondents' ages and their feelings toward Trump. A naive, but perhaps reasonable expectation would be that younger respondents have more negative (or "cold") feelings toward Trump. To get a sense of this, consider the simple regression using the `lm` function ("linear model") from base R. We store the model in object `reg_simple`. Once we have the model object saved, instead of using the `summary()` function from base R to display the results of the model, the `broom` package offers a "tidy" version of summarizing model objects in a cleaner, more robust way. Specifically, we will use the `tidy()`, `augment()`, and `glance()` functions from `broom` to explore our model in detail at both the variable and model levels.

```
reg_simple <- lm(fttrump ~ birthyr,
                data = NESdta_sub)
```

Before inspecting the output, notice that all of the analysis to this point follows the rules we laid out at the beginning – everything is an object and every action is a function. The `lm()` function is taking two arguments. The first is the `formula`, which has the dependent variable on the left, followed by a ~ ("approximately") symbol, and the independent variable on the right. The second is the `data` argument, which tells the function to which data to apply the formula. If you run `?lm` you can see what other arguments are available for this function.

The `lm()` function produces an object of class "lm" that we are saving to memory as `reg_simple`.

```
class(reg_simple)
```

```
## [1] "lm"
```

As we saw in the programming chapter, this object is also a list, which contains a number of other objects. If we use the `names()` function, we can see the names of these objects.

```
names(reg_simple)
```

```
##  [1] "coefficients"  "residuals"      "effects"
##  [4] "rank"          "fitted.values"  "assign"
##  [7] "qr"            "df.residual"    "xlevels"
## [10] "call"          "terms"          "model"
```

We can see a number of objects that we can access with the $ operator. For example, if we want to just call and retain the model coefficients, we can run the following code.

```
reg_simple$coefficients
```

```
## (Intercept)      birthyr
## 734.3474013   -0.3539436
```

Now, for a tidier version of the model results, we can call the `tidy()` function from the tidy-friendly `broom` package for a simple and clean description of the model output.

```
tidy(reg_simple)
```

```
## # A tibble: 2 x 5
##   term        estimate std.error statistic      p.value
##   <chr>          <dbl>     <dbl>     <dbl>        <dbl>
## 1 (Intercept)   734.      125.       5.87 0.00000000569
## 2 birthyr        -0.354    0.0636   -5.57 0.0000000321
```

The output includes `estimate`, `std.error`, `statistic`, and `p.value`. The `estimate` is the β coefficient, while the `std.error` is the measure of uncertainty surrounding that estimate. Then significance of this effect is captured by the `statistic` (usually either Z or t), as well as the `p.value`, which, despite the current controversy surrounding use and interpretation of p-values, is interpreted as the chance of observing some test statistic value equal to or more extreme than the computed value assuming the null hypothesis of no effect or relationship were true. To interpret our model, we start with the β coefficients, which are the effects we are estimating. In the simplest case, we interpret these values as *a one unit change in X causes a β change in Y*.

Next, `augment()` is another powerful function in `broom` that provides much more *variable*-level information useful for analysis. This function call returns verbose output including, e.g., fitted values (`.fitted`), residuals (`.resid`), Cook's distance (`.cooksd`), a measure of outliers discussed more below with diagnostics), and so on.

```
augment(reg_simple)
```

```
## # A tibble: 1,115 x 8
##    fttrump birthyr .fitted .resid .std.resid    .hat .sigma
##      <dbl>   <dbl>   <dbl>  <dbl>      <dbl>   <dbl>  <dbl>
## 1        1    1960    40.6  -39.6      -1.10 1.08e-3   36.0
## 2       28    1957    41.7  -13.7      -0.380 1.25e-3  36.1
## 3      100    1963    39.6   60.4       1.68 9.65e-4   36.0
## 4        0    1980    33.5  -33.5      -0.931 1.37e-3  36.0
## 5       61    1958    41.3   19.7       0.546 1.19e-3  36.1
## 6        5    1978    34.2  -29.2      -0.812 1.23e-3  36.0
## 7       85    1951    43.8   41.2       1.14 1.76e-3   36.0
## 8       70    1973    36.0   34.0       0.943 9.85e-4  36.0
```

```
## 9        5    1936   49.1  -44.1      -1.23  4.02e-3   36.0
## 10      74    1978   34.2   39.8       1.10  1.23e-3   36.0
## # ... with 1,105 more rows, and 1 more variable:
## #    .cooksd <dbl>
```

Finally, broom has another function, glance(), that returns *model*-level output, including R^2, log-likelihood values, AIC, BIC, degrees of freedom, and so on. See the output and inspect the package documentation for exhaustive details on the package and functions (i.e., ?broom).

```
glance(reg_simple)
```

```
## # A tibble: 1 x 12
##    r.squared adj.r.squared sigma statistic p.value      df
##        <dbl>         <dbl> <dbl>     <dbl>   <dbl>   <dbl>
## 1     0.0271        0.0262  36.0      31.0 3.21e-8       1
## # ... with 6 more variables: logLik <dbl>, AIC <dbl>,
## #    BIC <dbl>, deviance <dbl>, df.residual <int>,
## #    nobs <int>
```

Returning to the output, the negative β coefficient for birthyr suggests that younger respondents indeed have more negative feelings toward Trump. This is "significant" at the strict $p < 0.01$ level. We can also visualize our model by plotting it and overlaying a "best fit" line using ggplot() and adding a linear smoother layer (geom_smooth()) with confidence intervals around the line via se = TRUE. The result is in Figure 7.2.

```
ggplot(NESdta_sub, aes(x = birthyr, y = fttrump)) +
  geom_point(alpha = 0.7) +
  geom_smooth(method = "lm", se = TRUE, alpha = 0.1) +
  labs(x = "Birth Year",
       y = "Trump Feeling Thermometer Score",
       title = "The Effect of Age on Trump Feelings") +
  theme_minimal()
```

```
## `geom_smooth()` using formula 'y ~ x'
```

Now, we may worry about other effects, such as party affiliation, that may *also* influence feelings toward Trump. In such a case, we would want to update our model to account for this. This would be a multiple regression. To do so, simply add (yes, using the + operator) additional independent variables, or "regressors." In statistics, this is called "controlling" (e.g., "the effect of age, controlling for party").

We store the model in object reg_multiple, and then calculate predicted feelings toward Trump at the mean levels for each political party, while holding the other variable (birthyr) at its mean level using two powerful commands: tibble and predict.

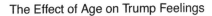

The Effect of Age on Trump Feelings

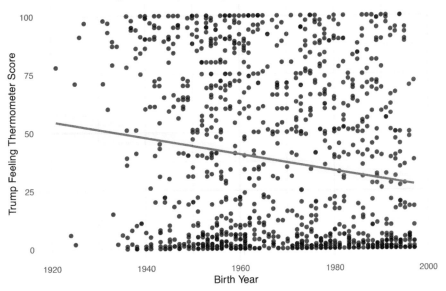

FIGURE 7.2

Linear Relationship between Age and Feelings toward Trump

```
reg_multiple <- lm(fttrump ~ birthyr + Party,
                   data = NESdta_sub); tidy(reg_multiple)
```

```
## # A tibble: 4 x 5
##    term              estimate std.error statistic  p.value
##    <chr>                <dbl>     <dbl>     <dbl>     <dbl>
## 1 (Intercept)          560.     108.         5.19 2.49e- 7
## 2 birthyr              -0.275     0.0548     -5.03 5.83e- 7
## 3 PartyIndependent      24.3      2.15       11.3  3.76e-28
## 4 PartyRepublican       46.4      2.36       19.7  4.23e-74
```

```
predict(reg_multiple,
    tibble(Party = c("Democrat", "Republican", "Independent"),
           birthyr = mean(NESdta_sub$birthyr)))
```

```
##        1        2        3
## 17.98660 64.33667 42.31621
```

As we would expect, feelings towards Trump are most positive among Republicans. Even as a candidate who had notably supported many Democrats in the past, Republicans appear to have been drawn to Trump to a much greater extent than Independents. We will come back to this point later in the chapter.

Exercises

7.5.0.0.1 Easy

- Why would you fit a regression model in the first place? What are the advantages compared to examining correlations, differences in means, or cross-tabulations?
- Create a similar regression model for support for Obama. Are the results different from what you saw in your model of support for Trump? If so how? If not, why do you think this is the case?

7.5.0.0.2 Intermediate

- How large is the gender gap in support for Trump? Add the `female` variable to your model to find out. Then use `predict()` to show the difference between men and women who are Republicans and of average age.
- How do you interpret a regression β coefficient?

7.5.0.0.3 Advanced

- Fit a regression model predicting support for Trump as a function of an interaction between political party affiliation and gender. What do you find, and what justification would there be to add a multiplicative interaction term to the right-hand side of a regression model?
- What do "ordinary", "least", and "squares" mean? Is "ordinary" problematic? Why or why not?

7.5.1 Regression Diagnostics

Upon fitting any model, researchers should always check the fit and diagnose their models. Though researchers may be interested in a variety of metrics for fit, the two most common checks for linear models are: *multicollinearity* (whether more than one independent variable is explaining roughly the same variance in the dependent variable) and *influential observations* (outliers exerting larger effect on the model fit over the other observations). There are a few methods for checking for these, but we will focus on two: variance inflation factor ("vif") for multicollinearity and Cook's distance for influential observations.

7.5.1.1 Multicollinearity

First, for multicollinearity, this is when we have multiple regressors explaining a lot of the same variance in our dependent variable. Recall the main goal of regression is to parsimoniously explain as much *unique* variance in the response variable as possible. When two explanatory variables are highly correlated, we encounter some amount of overlapping variance explained between the two.

This results in inefficiency, as the model is chewing up more degrees of freedom (working harder), but for relatively little (if any) additional explanatory gain. Multicollinearity can also result in misleading inference about the effect of particular variables. One of the most common tests to check for multicollinearity is to estimate variance inflation factor statistics for all variables in the model. Essentially, the test checks across every regressor in the full model, and then checks how much the variance of the model shifts when a variable is included versus when it is excluded. The simplest statistic for variance explained is the R^2. Thus, the formula is $1/1 - R_j^2$. Typically, values over 10 are considered problematic, though this is merely a rule of thumb, *not a statistical property.* The `vif()` function from the `car` package is quite simple, requiring only the model be supplied as input.

```
# First fit a multiple regression model
reg_full <- lm(fttrump ~ birthyr + Party + female,
               data = NESdta_sub); tidy(reg_full)
```

```
## # A tibble: 5 x 5
##    term            estimate std.error statistic  p.value
##    <chr>              <dbl>     <dbl>     <dbl>    <dbl>
## 1 (Intercept)       569.      108.         5.28 1.56e- 7
## 2 birthyr            -0.279     0.0547     -5.09 4.10e- 7
## 3 PartyIndependent   23.8       2.16       11.0  8.61e-27
## 4 PartyRepublican    46.1       2.36       19.6  1.24e-73
## 5 female             -3.94      1.87       -2.11 3.50e- 2
```

```
car::vif(reg_full)
```

```
##             GVIF Df GVIF^(1/(2*Df))
## birthyr 1.009648  1        1.004812
## Party   1.022983  2        1.005697
## female  1.015137  1        1.007540
```

We do not see any problematic variables based on VIF output. This is good news, suggesting the variables included are explaining unique variance in the dependent variable.[2]

7.5.1.2 Influential Observations and Outliers

Next, we can check for outliers that may be exerting a larger than expected amount of leverage or pull on the linear fit line explaining the data. We can

---

[2]Note, the lack of multicollinearity makes sense in such a simple case. The real threat of multicollinearity enters when there are many regressors included on the right-hand side of the model, and especially when the number of regressors approaches the size of the sample. There is a more complex statistical technique widely used in machine learning called regularization, which efficiently deals with multicollineary and model complexity when it is a serious threat. For interested readers, there are some excellent resources for learning these techniques in R (James et al., 2013).

check for these by calculating and visualizing Cook's Distance, which is one of the more common approaches to detecting outliers in regression models (as well as in data sets, though this application is not be covered in this chapter). Readers should note that you can inspect residual vs. fitted value plots in base R by simply plotting the lm object (i.e., plot(lm_model). However, here we will leverage two more recent packages from the *easystats* software group, which use ggplot2 from the tidyverse to render visualizations: performance to check for outliers, and see to plot the results. We will use our reg_full model previously fit.

Cook's distance calculates the influence of each observation on the fitted (predicted) values. It is a useful way to detect outliers, and whether any outliers may be troublesome for our estimates (i.e., pulling the regression fit line toward their location in the predictor space). Yet, whether the observation is a *problematic* outlier is a question left to the researcher. Let's check for outliers in our reg_full model using the check_outliers() function from performance.

```
check_outliers(reg_full)
```

```
## OK: No outliers detected.
```

Good news: no outliers were detected, at least when using Cook's distance. Note: users should inspect the performance package documentation for the list of available metrics, which can be included and changed by supplying the appropriate name (e.g., cook or iforest) to the method argument in the check_outliers() function.

Yet, while this is good news for our model, it is not such good news when demonstrating what to do if and when outliers are detected. To make this point, and demonstrate the available tools, we will replicate the example from the performance package documentation, to which Waggoner contributed. This example is using the mtcars data set, but with fake outlier cases manually added to the mt2 data set. Upon creating the data with the outliers, we will use the check_outliers() function previously used, and then plot the results using the plot() function from the see package. The result will be an easy to read ggplot2 object with outliers labeled accordingly in Figure 7.3.

```
# create the synthetic data with outliers
mt1 <- mtcars[, c(1, 3, 4)]
mt2 <- rbind(mt1, data.frame(mpg = c(37, 40),
                             disp = c(300, 400),
                             hp = c(110, 120)))

# fit the model on the created data
model <- lm(disp ~ mpg + hp,
            data = mt2)
```

```
# check for outliers using Cook's and IQR (for comparison)
check_outliers(model, method = c("cook", "iqr"))
```

```
## Warning: 3 outliers detected (cases 31, 33, 34).
```

```
# visualize
plot(check_outliers(model, method = c("cook", "iqr"))) +
  theme(axis.text.x = element_text(angle = 75, hjust = 1))
```

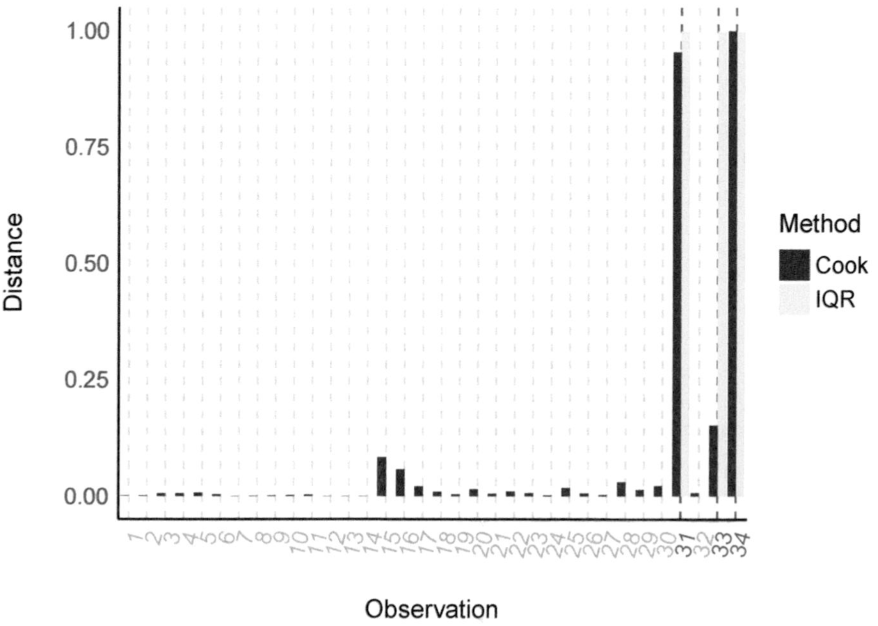

FIGURE 7.3
Labeled Outliers via Cook's Distance and IQR

Exercises

7.5.1.2.1   Easy

- Run these diagnostics on your regression for support for Obama from the previous section. Are there any outliers or issues you notice in this regression?

7.5.1.2.2   Intermediate

- What is the first step you would take if you suspected an outlier may be exerting a large amount of influence on the fit of your model?
- Replicate the previous case, but this time change `method = "all"`. Do

you see consistency across the different metrics? Why or why not, do you think?

7.5.1.2.3 *Advanced*

- What is in the denominator of the variance inflation factor equation, and why is this the case?
- Think about the logic behind Cook's distance. Now, look up the "local outlier factor" (*LOF*). How do these differ in substantive terms? How are they similar?

7.5.2 Saving Regression Results

Even if you follow best practices and present your results visually, you will likely need to provide a table of your regression results at some point. Here again, the `stargazer` package is useful for automatically generating these tables.

Here is how we would create a table for Microsoft Word of the two regression models above. You will notice that this code is very similar to what we used to create the table of summary statistics. The only real difference is that we are including more than one object from which `stargazer()` is drawing information. We also need to set the label for the dependent variable (`dep.var.labels()`) to make it more informative than `fttrump` (and similarly for the independent variables via `covariate.labels` argument). The output, when we open it in Word, is a publication-ready table.

```
stargazer(reg_simple, reg_multiple,
          dep.var.labels = c("Approval of Trump"),
          covariate.labels = c("Birth Year",
                               "Independent",
                               "Republican"),
          type = "html",
          out = here("tables", "ols_models.doc"))
```

As we noted before, there are a number of different options with `stargazer`, so take some time to play around with these to find your favorite table format.

Exercises

7.5.2.0.1 *Easy*

- Create a table that includes a bivariate model and a multivariate model of support for Obama (`ftobama`). Save it and open it in Microsoft Word or another word processing program.

- You can add a number of different models to the same table. Take the table for approval of Trump (`fttrump`) above and add the bivariate and multivariate models for approval of Obama (`ftobama`). *Be sure to modify the labels accordingly.*

- What is the difference between `kable()` and `stargazer()`? (*hint*: consider looking into the `knitr` package in the Tidyverse)

7.5.3   Concluding Remarks for OLS

We have merely scratched the surface on fitting, interpreting, and diagnosing linear models. For example, you can use regressions to fit mediation models when mediating effects are suspected (Waggoner, 2020). There are also many other diagnostic tests you could (and should) run when you fit models and present results, such as studentized residual plots, leverage plots, and so on. Many of these other techniques are detailed in the easystats `performance` package demonstrated above.

The bottom line is, in social science research, researchers should always strive to be honest and thorough in the research program and present the full scope of the process. This includes multiple iterations of models run, diagnostic tests, and even alternative specifications. And more specifically for our purposes, we suggest the Tidyverse is an exceptionally useful environment to facilitate this process in a consistent, clean manner.

7.6   Binary Response Models

Recall that if we are interested in predicting the outcome of a binary dependent variable (e.g., moving from a no to a yes, or the probability of moving from a 0 to a 1), then we should fit a binary response model that can efficiently handle estimation of the outcome.

OLS is inappropriate for binary response dependent variables, because it assumes a continuous distribution in the response. The first attempt to deal with this type of data was called a linear probability model (LPM). But, it was soon realized that an LPM also produced unrealistic probabilities (e.g., 105% or -30% likelihood of some event happening).

As such, the two most widely used models aimed at handling these types of data in social sciences are logistic regression (logit) and probit regression. Probit was much more popular in the 1980s and 1990s. Today, however, logistic regressions are arguably used more frequently in the social sciences. It is important to note that this preference is largely cosmetic (at least in a substantive, inferential sense), as both estimators produce virtually identical point estimates. We will demonstrate this below.

Ultimately, though, fitting a binary response model in R is nearly as straightforward as fitting a basic linear model. This time, though, we will use the `glm()` function instead of `lm()`, as logistic and probit regressions are *generalized* linear models (hence the "g" in the "glm" function). Once we fit our model, it is always a good idea to visualize the results as well as check for the robustness of our estimates. We demonstrate these concepts in the Tidyverse for the remainder of this chapter.

7.6.1 Loading Some New Libraries

First, we need to load a few new libraries and then create a binary response variable. To do the latter, we will call it `pro_trump`, where over 50% on the Trump feeling thermometer suggests the respondent supports Trump, at least more than opposing him. To create this new variable, we use the `ifelse()` function you learned earlier. Ultimately, we are interested in predicting the likelihood of supporting Trump, relative to *not* supporting him. In other words, we are interested in the probability of moving from a 0 (not support) to a 1 (support).

```
# load some packages/libraries first
library(faraway)
library(foreign)
library(ggplot2)
library(arm)
library(MASS)
library(OOmisc) # for ePCP fit statistics
library(pROC) # for plotting ROC curves
library(lmtest) # for likelihood ratio tests
library(skimr)

# create new "pro_trump" var for prediction
NESdta_sub <- NESdta_sub %>%
  mutate(pro_trump = ifelse(fttrump >= 50, 1, 0)) %>%
  drop_na()

# inspect to make sure everything looks right
sample_n(tibble(NESdta_sub$pro_trump), 5)

## # A tibble: 5 x 1
##   `NESdta_sub$pro_trump`
```

```
##                      <dbl>
## 1                        1
## 2                        1
## 3                        0
## 4                        1
## 5                        0
```

```
table(NESdta_sub$pro_trump) # whole df: 665 = 0; 450 = 1
```

```
##
##   0   1
## 665 450
```

7.6.2   Demonstrating Why OLS is Poor for Binary Outcomes

First, to motivate the value of fitting a logistic or probit model, we demonstrate how an OLS model performs poorly in predicting binary responses in Figure 7.4.

```
ggplot(NESdta_sub, aes(x = birthyr, y = pro_trump)) +
  geom_point(alpha = 0.7) +
  geom_smooth(method = "lm", se = TRUE) +
  labs(x = "Birth Year",
       y = "Observed Pro-Trump Rating (FT >= 0.5)",
       title = "The Effect of Age on Pro-Trump Rating") +
  theme_minimal()
```

```
## `geom_smooth()` using formula 'y ~ x'
```

The data constrained at $\{0, 1\}$, we can see the linear fit line is quite inefficient and does not explain very much of the data. Thus we need a model that can handle binary response dependent variables. As previously noted, there are two options here that are most commonly used: logit and probit. However, which one should we choose? The short answer is, *it doesn't really matter*. But let's prove it Tidyverse style!

7.6.3   Demonstrating Logit and Probit are (Virtually) Identical

To demonstrate that logit and probit are functionally identical, we first fit a logit model, and then a probit model to estimate the relative impact of respondents' ages (`birthyr`) on the likelihood of being "pro-Trump". Of note,

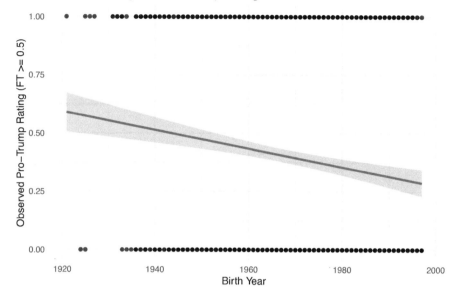

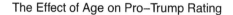

FIGURE 7.4
Poor Fit of OLS for a Binary Outcome

the only thing we are changing in these models is the link function, from `logit` to `probit`. We store each model in objects `logit` and `probit`.[3]

```
logit <- glm(pro_trump ~ birthyr,
             family = binomial(link = logit),
             NESdta_sub); tidy(logit)
```

```
## # A tibble: 2 x 5
##   term         estimate std.error statistic   p.value
##   <chr>           <dbl>     <dbl>     <dbl>     <dbl>
## 1 (Intercept)     32.5      7.20       4.52 0.00000622
## 2 birthyr         -0.0167   0.00366   -4.57 0.00000483
```

```
probit <- glm(pro_trump ~ birthyr,
              family = binomial(link = probit),
              NESdta_sub); tidy(probit)
```

```
## # A tibble: 2 x 5
##   term         estimate std.error statistic   p.value
```

---

[3]The update to the `glm()` function compared to the `lm()` function is the inclusion of `family` argument. This is where we tell the function that we are interested in the binomial family, and that we want either a logit or a probit link *within the binomial family*.

```
##    <chr>           <dbl>      <dbl>       <dbl>       <dbl>
## 1 (Intercept)     20.3       4.44        4.56 0.00000508
## 2 birthyr        -0.0104     0.00226    -4.62 0.00000391
```

Importantly, raw coefficients from logit and probit models are not extremely helpful beyond direction of effects and significance. To turn these into something more useful, a common choice is to calculate predicted probabilities. To do so, and thus compare both numerically and visually, we can show the predicted probabilities for being pro-Trump for specific levels of age. We will show for the oldest respondent (born in 1921), the median respondent (born in 1967) and then the youngest respondent (born in 1997). We obtain these values using skim(). We then calculate and store the predicted values at each level by simply plugging the intercept (β_0) and slope (β_j) coefficients into either ilogit() or pnorm() for logit and probit models, respectively. The reason for this choice is because the logit requires the inverse logistic distribution, while the probit requires the normal distribution to turn these coefficient values into predicted probabilities for more intuitive interpretation. We then store these in a transposed tibble using the tribble() function to offer a cleaner look at predicted probabilities by age level.

```
# get the different values for min, med, and max birth year first
summary(NESdta_sub$birthyr)

# store preds when birth year is at its min, median, and max
l_min <- ilogit(32.52570 + (-0.01673) * 1921) #min
p_min <- pnorm(20.263235 + (-0.010424) * 1921) #min
l_med <- ilogit(32.52570 + (-0.01673) * 1967) #median
p_med <- pnorm(20.263235 + (-0.010424) * 1967) #median
l_max <- ilogit(32.52570 + (-0.01673) * 1997) #max
p_max <- pnorm(20.263235 + (-0.010424) * 1997) #max

Predictions_Logit_Probit <- tribble( # transposed tibble
  ~` `, ~Logit, ~Probit,
  "Minimum Birth Year (1921)", l_min, p_min,
  "Median Birth Year (1967)", l_med, p_med,
  "Maximum Birth Year (1997)", l_max, p_max
)
Predictions_Logit_Probit
```

In line with the earlier OLS findings, younger respondents are much less likely to be in the pro-Trump camp, while older respondents are most likely to be in the pro-Trump camp. But more importantly for our purposes, note the virtually identical predictions for both the logit and probit models. This is evidence of point number 1. For evidence of point number 2, corroborating these similarities, we can also visualize this by plotting the predicted probabilities against each other. If they are the same, then we would expect a perfectly diagonal 45 degree line from the lower left to the upper right. To do this, we will use ggplot2 with a point geometry to create a scatterplot and present the comparison in Figure 7.5.

```
logit_phat <- logit$fitted.values # fitted values from logit
probit_phat <- probit$fitted.values # fitted values from probit

hat_data <- tibble(logit_phat, probit_phat)
#hat_data # uncomment to inspect if you'd like

hat_data %>%
  ggplot() +
  geom_point(alpha = 0.7, aes(x = logit_phat, y = probit_phat)) +
  labs(x = "Logit Predicted Probabilities",
       y = "Probit Predicted Probabilities",
       title = "Comparing Logit & Probit Predictions") +
  theme_minimal()
```

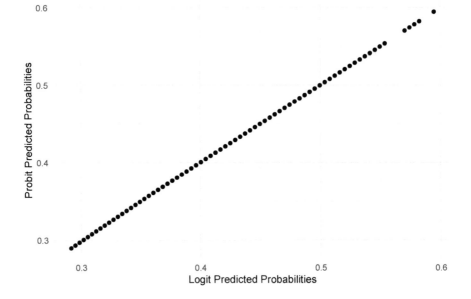

FIGURE 7.5
Comparing Logit and Probit

It is clear from this chart that the logit and probit are virtually identical. Thus, we will proceed with only logit for the remainder of the chapter.

7.6.4 Hypothesis Testing, Inference, and Substantive Interpretation

For both the logit and probit models, the glm() function returns the slope coefficients, their corresponding standard errors and significance levels. We can also get confidence intervals (CIs) for the estimated coefficients using the confint() function, which profiles the likelihood distribution. After fitting the model, we can convert coefficients into odds ratios, which are straightforward to interpret. Based on the mean odds ratio, a one-point increase in X, will increase the probability of moving from 0 to 1 by a factor of Z. Values greater than 1 are positive relative effects, whereas values less than 1 are negative relative effects. Note that the odds ratios are the exponentiated coefficients from the model, and can be calculated via the exp() function from base R.

```
confint(logit)
```

```
## Waiting for profiling to be done...
```

```
##                    2.5 %        97.5 %
## (Intercept) 18.48486848 46.717426131
## birthyr     -0.02394826 -0.009594217
```

```
base::exp(logit$coefficients)
```

```
##  (Intercept)         birthyr
## 1.335769e+14 9.834070e-01
```

7.6.5 A Multivariate Model

To this point, we have found that age has a significantly negative impact on the likelihood of being pro-Trump. Yet, there are likely other factors that also matter. To explore these, and thus control for other factors, we can complicate our base model by adding additional regressors as we did in the OLS case earlier in the chapter.

```
mult_logit <- glm(pro_trump ~ birthyr + factor(Party) + gender,
                family = binomial(link = logit),
                NESdta_sub); tidy(mult_logit)
```

```
## # A tibble: 5 x 5
##   term                  estimate std.error statistic  p.value
##   <chr>                    <dbl>     <dbl>     <dbl>    <dbl>
## 1 (Intercept)            31.1       8.06       3.86 1.16e- 4
## 2 birthyr                -0.0164    0.00409   -4.00 6.44e- 5
## 3 factor(Party)Indepe~    1.32      0.163      8.11 5.13e-16
## 4 factor(Party)Republ~    2.41      0.182     13.3  3.14e-40
## 5 gender                 -0.275     0.138     -2.00 4.56e- 2
```

Here we store the multivariate version in the object `mult_logit`, and also make `Party` a factor for the sake of plotting below. Note that this does not change the impact of the variable or the model.

With a more fully specified model controlling for additional factors (party and also gender), we can get a more reliable sense of the magnitude of these effects by generating out-of-sample predicted probabilities, ranging over the birth year and holding the effect of gender at its mean value. This will give us a targeted look at the effect of party affiliation on the likelihood of being pro-Trump.

```
# out of sample predicted values
# let birth year range (do this 300 time for each level of party)
# hold gender effect at mean; 100 times for each party id level
sub_data <- with(NESdta_sub, tibble(
  birthyr = rep(seq(from = 1921, to = 1997,
                    length.out = 100),
                3),
  gender = mean(gender),
  Party = factor(rep(c("Democrat",
                       "Republican",
                       "Independent"),
                     each = 100)))
)


# combine predicted values and SEs based on "sub_data"
pred_data <- cbind(sub_data, predict(mult_logit,
                                     newdata = sub_data,
                                     type = "link",
                                     se = TRUE))


# store lower limit (LL) and upper limit (UL) values
# attach to predicted values data frame created in "pred_data"
pred_data <- within(pred_data, {
  PredictedProb <- plogis(fit)
  LL <- plogis(fit - (1.96 * se.fit))
  UL <- plogis(fit + (1.96 * se.fit))
})
```

With our synthetic out-of-sample data frame created, we can now plot these results with unique lines and confidence intervals for each party in Figure 7.6.

```
ggplot(pred_data, aes(x = birthyr, y = PredictedProb)) +
  geom_errorbar(aes(ymin = LL, ymax = UL), alpha = 0.2) +
  geom_line(aes(color = Party), size = 1) +
  scale_color_manual(values = amerika_palette("Dem_Ind_Rep3"),
                     name = "Party") +
```

```
labs(x = "Birth Year",
     y = "Predicted Probability of Pro-Trump Rating",
     title = "The Effect of Age and Party on Pro-Trump Rating",
     subtitle = "Trends from 300 Out-of-Sample Predictions") +
theme_minimal()
```

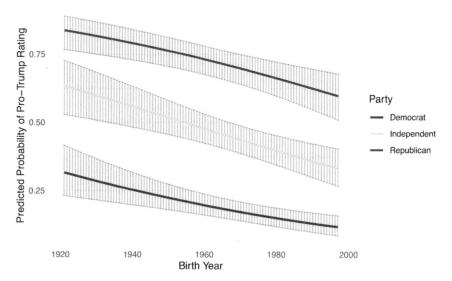

FIGURE 7.6
Out-of-Sample Predictions

In line with expectations and seen in the different slopes for each level of party
affiliation, Democrats have the overall lowest probability of supporting Trump,
followed by Independents in the middle, and followed by Republicans, who
have the highest overall probability of supporting Trump. Yet, the likelihood
of being pro-Trump across all parties drastically decreases as the respondent
pool gets younger.

7.6.6 Assessing Model Fit

As with OLS regression, it is vitally important to assess the fit of models in
the binary response world as well. We focus on two difference approaches:
classification-based (did the model classify observations correctly compared
to true values) and the likelihood-based (does model X predict the likelihood of moving from 0 to 1 better than model Z?). We start with expected

proportion correctly predicted (ePCP) and then we inspect receiver operating characteristic (ROC) curves. We conclude with likelihood ratio tests.

7.6.6.1   Expected Proportion Correctly Predicted (ePCP)

First, using function ePCP(), we can calculate the expected proportion correctly predicted (ePCP) statistics associated with each of our logit models (the bivariate and the multivariate). We then store the predicted values and present them visually across both models. Here, we are interested in which model performs "best." Higher ePCP suggests a better fit, or a higher proportion of correctly classifying Trump supporters versus non-supporters.[4]

```
y <- NESdta_sub$pro_trump
pred1 <- predict(logit, type="response")
pred2 <- predict(mult_logit, type="response")

epcp1 <- ePCP(pred1, y, alpha = 0.05)
epcp2 <- ePCP(pred2, y, alpha = 0.05)
```

The multivariate iteration has a higher mean ePCP value than the bivariate model, suggesting the more complicated multivariate model fits the data better than the bivariate model. We can also visualize these results in Figure 7.7.

```
epcpdata <- data.frame(rbind(epcp1, epcp2))
epcpdata$model <- c(1,2)
epcpdata$count <- factor(c(1,2),
                         label = c("Bivariate", "Multivariate"))

ggplot(epcpdata, aes(x = model, y = ePCP,
                     color = count)) +
  geom_bar(position = position_dodge(),
           stat = "identity",
           fill = "darkgray") +
  geom_errorbar(aes(ymin = lower, ymax = upper),
                width = 0.1,
                position = position_dodge(0.9)) +
  labs(title = "Comparing ePCP between Bivariate and\n
       Multivariate Logistic Regressions",
       x = "Model Specification",
       y = "Expected Proportion of Correct Prediction",
       color = "Model") +
  theme_minimal()
```

---

[4]We thank Ling Zhu (University of Houston) for sharing some excellent base code used in these assessment tests.

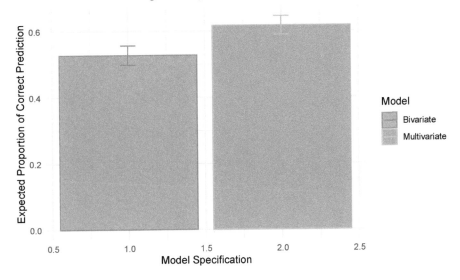

FIGURE 7.7
Expected Proportion Correctly Predicted (ePCP)

7.6.6.2 Receiver Operating Characteristic (ROC) Curves

Next, receiver operating characteristic (ROC) curves plot the correct predictions (sensitivity, "true positive" rate) against false predictions (specificity, "false positive" rate). When a model fits well, the area under the curve (AUC) will be greater, where 1 suggests perfect classification. The 45-degree diagonal line is a reference point, such that we are interested in the model with the curve most distant to the upper left from the diagonal line, suggesting greater, positive AUC, and thus a better fit with more true positives correctly classified, which again is support for Trump ($y = 1$). Results are in Figure 7.8.

```
par(mfrow = c(1,2)) # set the pane side by side (rows, columns)
plot.roc(y, pred1,
         col="darkgreen",
         main = "Bivariate Logit")

## Setting levels: control = 0, case = 1

## Setting direction: controls < cases

plot.roc(y, pred2,
         col="darkorange",
         main = "Multivariate Logit")
```

```
## Setting levels: control = 0, case = 1
## Setting direction: controls < cases
```

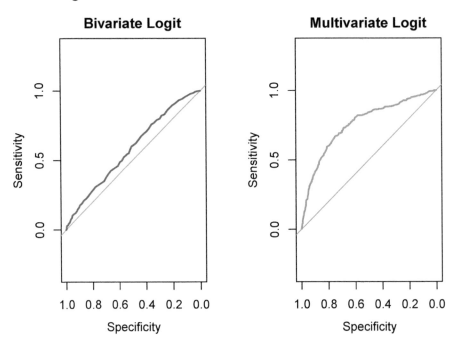

FIGURE 7.8
ROC Curves Comparing Bivariate and Multivariate Fits

Similar to ePCP, we can see much greater AUC for the multivariate model with a curve farther to the upper left compared to the bivariate model, suggesting the multivariate specification fits best.

7.6.6.3   Likelihood Ratio Tests

Finally, we can also assess fit by comparing the fit between models based on the likelihood ratios using the likelihood ratio test. The test statistic is defined as, $LR_{test} = 2lnL(M_B) - 2lnL(M_M)$, where $L(M_B)$ is the likelihood of estimates for the bivariate model and $L(M_M)$ is the likelihood of estimates for the multivariate model.

```
lrtest(logit, mult_logit)
```

```
## Likelihood ratio test
##
## Model 1: pro_trump ~ birthyr
## Model 2: pro_trump ~ birthyr + factor(Party) + gender
##    #Df  LogLik Df  Chisq Pr(>Chisq)
```

```
## 1    2 -741.35
## 2    5 -632.56  3 217.58  < 2.2e-16 ***
## ---
## Signif. codes:
## 0 '***' 0.001 '**' 0.01 '*' 0.05 '.' 0.1 ' ' 1
```

As the log-likelihood of a model is a measure of fit, we are looking for a significant result (sufficiently small p-value), and the model with a smaller absolute log-likelihood value. Seen from the `lrtest()` output, the multivariate model is indeed better fitting than the bivariate model, in line with the previous tests.

Exercises

7.6.6.3.1 Easy

- Repeat the modeling exercise above, but with a dichotomous version of the feeling thermometer for Obama (`ftobama`). How are the results similar or different?
- Create a table for your logit models of support for Trump and Obama.

7.6.6.3.2 Intermediate

- Check the fit of your Obama logistic regression. Do the independent variables contribute to a better model for explaining support for Obama or support for Trump? How do you know? Why do you think these results are different, if indeed they are?
- Suppose you visualized an ROC curve for your logistic regression model, but the curve was to the lower right of the 45 degree reference line (i.e., below). What would this tell you, and how would you know?

7.6.6.3.3 Advanced

- What might be a reasonable argument against collapsing a quantitative (i.e., continuous numeric) variable into a binary variable?
- Suppose your teacher said, "fit a linear probability model with a binary dependent variable." What would you say in response and what would be a different approach you could take and why?

7.7   Concluding Remarks

This chapter has attempted to cover a lot of ground in a very short amount of space. Our goal was to demonstrate that the range of statistical tools common to most social scientists fits quite well in the Tidyverse. For example, those familiar with the base R function for correlation, `cor()`, will immediately see that working with the Tidyverse counterpart, `correlate()`, is much easier.

The capacities of R for statistical modeling are immense. From Bayesian analysis to machine learning, you can find just about any type of statistical model you will need, with an R package already developed. If you are wondering where to start to find a specific model, visit the CRAN Task Views. Consider also searching for the model or technique, adding "in R" to the search. Such a task with return a host of tutorials, blog posts, and many other resources to help you along your way.

8

Parting Thoughts

This book has provided a very concise introduction to R and the Tidyverse. We hope we have made a sufficient case for the use of both, and provided you with the tools and understanding you need to set you off on your journey. But you may be wondering where you go from here. In this section, we will provide you with a few ideas of how to move towards mastery of the R language, and get lots of great ideas for how you can use R to create new, original, and exciting research. As we said in the beginning, learning to program is not always an easy process. But over time, you will find these tools will open areas of research that you never thought of prior to learning programming. As the old saying goes,

If all you have is a hammer, everything begins to look like a nail.

Learning to efficiently program in R will change the way you view and do your research. Neither of us anticipated the kind of work we are doing now when we started graduate school. Yet, learning R opened new avenues far more exciting than anything we had originally anticipated.

8.1 Continuing to Learn with R

As you leave this book, one thing is more important than any other for you to learn R – use R. You probably have heard of other tools for research that offer a simpler (at least at first) way to accomplish what you want to do. SPSS and Stata have dropdown menus – why not do basic analyses there? The answer is that you will not learn R if you are only using it every once in a while, when you need to do something you cannot do in another language. We will not go as far as one well-known scholar who claimed to do his taxes in R, but you will not really learn without continuous use. Try to use R as your first choice for analysis, and only use another program if you find yourself in a situation in which it is really needed. R should become your default. This is part of the reason we emphasized data management and graphics in this book. Since these are the tasks that begin just about any project, you have no excuse not to start with R.

As you program with R, you will build a base of code that you will continue to use as you work. Remember to save the scripts that you write. You will find that you can re-use your code over and over. And, as you develop a base of scripts, you will find that working in R becomes much faster (and faster than using drop-down menus all the time). The online companion site for this book provides all the code from this book to jump-start this process, and you can find a range of code available online to help you build up this base of code.

Related to this, it is a good idea to subscribe to daily emails from R-Bloggers. This will give you exposure to many of the exciting projects that are being done by others in R, and will allow you to see the many opportunities using R opens to your research. Many places also have dedicated groups for using R or data science more generally, where you can meet other R users and participate in fascinating projects, no matter your level of skill. You can find many of these on Meetup.

Finally, be patient with yourself. There is the old story (perhaps apocryphal) that Einstein told a student, who claimed to have difficulty with math, "Do not worry about your difficulties in mathematics. I can assure you mine are still greater." All of us have had situations where we have struggled to get a particular piece of code to run correctly, or have received an error message we do not understand. Keep working on it and looking for help. It may take a while, but there is no greater feeling than conquering, and mastering, a task that you have struggled with previously. Celebrate your accomplishments, and persist through your difficulties.

8.2   Where To Go from Here

As we have mentioned in several places, our online companion site provides code examples of several other common (and some uncommon) tasks in R. You can download these and add them to your code base.

To discover specific packages in R that are useful for a particular statistical model or task, there is also the CRAN Task Views, which provides a curated list of packages available for all kinds of analysis, from Bayesian statistics to network analysis and machine learning.

Many scholars have also put together books that will help you as you work with R in more specific circumstances, and many of these are available online (many at no cost). A compendium of these books can be found at https://www.r-project.org/doc/bib/R-books.html.

A few more specific books that you might pick up after this book are:

– Hadley Wickham and Garrett Grolemund's *R for Data Science*. This book provides a more comprehensive picture of what you can do in the Tidyverse. It is much more valuable, however, once you already have some basic familiarity working with R and the basics of the Tidyverse, as it is written for an audience with some level of prior programming experience. The book is available in print, as an electronic book, or online for free (https://r4ds.had.co.nz/).

– Quan Li's *Using R for Data Analysis in Social Sciences.* Li provides an excellent introduction to base R, and also goes into much greater detail on specific statistical models and, in particular, how to replicate studies in the social sciences.

– Kieran Healy's *Data Visualization: A Practical Introduction.* This book provides an overview of the graphical capabilities in R using practical examples using `ggplot2`.

– For those interested in Bayesian statistics using R, we cannot recommend Richard McElreath's *Statistical Rethinking* highly enough. It is both entertaining and enlightening, and uses R to demonstrate important statistical concepts with which any researcher should be familiar.

8.3   A Final Word

In so many ways, we are living in a golden era for quantitative social science research. Never have we had so much data available on human behavior, and the ability to generate and analyze these data in ways that would have been inconceivable just a decade ago. The future of the social sciences belongs to those who are able to produce unique and replicable research. You now have at your disposal what we consider one of the most powerful tools for achieving this. We look forward to seeing what you do with it.

Bibliography

Aldrich, J. H. and McGraw, K. M. (2012). *Improving public opinion surveys: interdisciplinary innovation and the american national election studies.* Princeton University Press.

Börner, K., Bueckle, A., and Ginda, M. (2019). Data visualization literacy: Definitions, conceptual frameworks, exercises, and assessments. *Proceedings of the National Academy of Sciences*, 116(6):1857–1864.

Campbell, A., Converse, P. E., Miller, W. E., and Stokes, D. E. (1960). *The american voter.* University of Chicago Press.

Collaboration, O. S. et al. (2015). Estimating the reproducibility of psychological science. *Science*, 349(6251):aac4716.

Faraway, J. J. (2016). *Extending the linear model with R: generalized linear, mixed effects and nonparametric regression models.* Chapman and Hall/CRC.

Finlay, B. and Agresti, A. (1986). *Statistical methods for the social sciences.* Dellen.

Fox, J. and Weisberg, S. (2018). *An R companion to applied regression.* Sage Publications.

Freese, J. and Peterson, D. (2017). Replication in social science. *Annual Review of Sociology*, 43:147–165.

Gailmard, S. (2014). *Statistical modeling and inference for social science.* Cambridge University Press.

Gelman, A. and Hill, J. (2006). *Data analysis using regression and multilevel/hierarchical models.* Cambridge university press.

Giani, M. and Méon, P.-G. (2017). Global racist contagion following donald trump's election. *British Journal of Political Science*, pages 1–8.

Healy, K. (2018). *Data Visualization: A Practical Introduction.* Princeton University, Princeton, NJ.

Hlavac, M. (2016). Extremebounds: Extreme bounds analysis in r. *Journal of Statistical Software*, 72.

Hlavac, M. (2018). *stargazer: Well-Formatted Regression and Summary Statistics Tables.* R package version 5.2.2.

Ihaka, R. and Gentleman, R. (1996). R: a language for data analysis and graphics. *Journal of computational and graphical statistics*, 5(3):299–314.

Ikenberry, G. J. (2017). The plot against american foreign policy: Can the liberal order survive. *Foreign Aff.*, 96:2.

Inglehart, R. and Welzel, C. (2009). How development leads to democracy: What we know about modernization. *Foreign Affairs*, pages 33–48.

James, G., Witten, D., Hastie, T., and Tibshirani, R. (2013). *An introduction to statistical learning*, volume 112. Springer.

Jones, L. V. (1987). *The Collected Works of John W. Tukey: Philosophy and Principles of Data Analysis 1965-1986*, volume 4. CRC Press.

Kennedy, R. (2010). The contradiction of modernization: A conditional model of endogenous democratization. *The Journal of Politics*, 72(3):785–798.

Kennedy, R. and Tiede, L. (2013). Economic development assumptions and the elusive curse of oil. *International Studies Quarterly*, 57(4):760–771.

King, G. (1995). Replication, replication. *PS: Political Science & Politics*, 28(3):444–452.

Kruger, J. and Dunning, D. (1999). Unskilled and unaware of it: how difficulties in recognizing one's own incompetence lead to inflated self-assessments. *Journal of personality and social psychology*, 77(6):1121.

Leamer, E. E. (1983). Let's take the con out of econometrics. *The American Economic Review*, 73(1):31–43.

Leamer, E. E. (2010). Tantalus on the road to asymptopia. *Journal of Economic Perspectives*, 24(2):31–46.

Leemis, L. (2016). *Learning Base R.* Lightning Source.

Lerner, D. (1958). The passing of traditional society: Modernizing the middle east.

Levine, R. and Renelt, D. (1992). A sensitivity analysis of cross-country growth regressions. *The American economic review*, pages 942–963.

Lewis-Beck, M. S., Jacoby, W. G., Norpoth, H., and Weisberg, H. F. (2008). *The American voter revisited.* University of Michigan Press.

Li, Q. (2018). *Using R for Data Analysis in Social Sciences.* Oxford University, Oxford, UK.

Lipset, S. M. (1959). Political man: The social bases of politics.

Lüdecke, D., Makowski, D., and Waggoner, P. D. (2019). Performance: assessment of regression models performance. *R package version 0.4*, 2.

Lüdecke, D., Makowski, D., Waggoner, P. D., and Ben-Shachar, M. S. (2020). see: Visualisation toolbox for 'easystats' and extra geoms, themes and color palettes for 'ggplot2'. *R package version 0.5.1.1*.

MacWilliams, M. C. (2016). Who decides when the party doesn't? authoritarian voters and the rise of donald trump. *PS: Political Science & Politics*, 49(4):716–721.

Matloff, N. (2011). *The art of R programming: A tour of statistical software design*. No Starch Press.

McAleer, M., Pagan, A. R., and Volker, P. A. (1985). What will take the con out of econometrics? *The American Economic Review*, 75(3):293–307.

McNamara, A., Arino de la Rubia, E., Zhu, H., Ellis, S., and Quinn, M. (2019). *skimr: Compact and Flexible Summaries of Data*. R package version 1.0.5.

Monogan III, J. E. E. (2015). *Political Analysis Using R*. Use R! Springer, New York.

Oakley, B. A. (2014). *A mind for numbers: How to excel at math and science (even if you flunked algebra)*. TarcherPerigee.

O'donnell, G. (1973). Modernization and bureaucratic-authoritarianism: Studies in south american politics.

Pierson, P. (2017). American hybrid: Donald trump and the strange merger of populism and plutocracy. *The British journal of sociology*, 68:S105–S119.

Radford, J. and Lazer, D. (2019). Big data for sociological research. *The Wiley Blackwell Companion to Sociology*, pages 417–443.

Robinson, D. (2014). broom: An r package for converting statistical analysis objects into tidy data frames. *arXiv preprint arXiv:1412.3565*.

Robinson, J. A. (2006). Economic development and democracy. *Annu. Rev. Polit. Sci.*, 9:503–527.

RStudio Team (2015). *RStudio: Integrated Development Environment for R*. RStudio, Inc., Boston, MA.

Sides, J., Tesler, M., and Vavreck, L. (2017). The 2016 us election: How trump lost and won. *Journal of Democracy*, 28(2):34–44.

Smith, T. (1979). The underdevelopment of development literature: the case of dependency theory. *World Politics*, 31(2):247–288.

Tufte, E. R. (2001). *The visual display of quantitative information*, volume 2. Graphics press Cheshire, CT.

Tufte, E. R., Goeler, N. H., and Benson, R. (1990). *Envisioning information*, volume 126. Graphics press Cheshire, CT.

Venables, W. N. and Ripley, B. D. (2013). *Modern applied statistics with S-PLUS*. Springer Science & Business Media.

Waggoner, P. D. (2018a). *Advice to Young (and Old) Programmers: A Conversation with Hadley Wickham*. R-Bloggers.

Waggoner, P. D. (2018b). The hhi package: Streamlined calculation and visualization of herfindahl-hirschman index scores. *Journal of Open Source Software*, 3(28):828.

Waggoner, P. D. (2019). *amerika: American Politics-Inspired Color Palette Generator*. R package version 0.1.0.

Waggoner, P. D. (2020). A simple method for purging mediation effects. *Journal of Statistical Theory and Practice*, 14(25):25.

Ward, M. D. and Gleditsch, K. S. (2018). *Spatial regression models*, volume 155. Sage Publications.

Wickham, H. (2009). *ggplot2: Elegant Graphics for Data Analysis*. Springer, New York.

Wickham, H. (2014). Tidy data. *Journal of Statistical Software*, 59(10):1–23.

Wickham, H. (2017). *tidyverse: Easily Install and Load the 'Tidyverse'*. R package version 1.2.1.

Wickham, H., Averick, M., Bryan, J., Chang, W., McGowan, L. D., François, R., Grolemund, G., Hayes, A., Henry, L., Hester, J., et al. (2019a). Welcome to the tidyverse. *Journal of Open Source Software*, 4(43):1686.

Wickham, H., Chang, W., Henry, L., Pedersen, T. L., Takahashi, K., Wilke, C., and Woo, K. (2019b). *ggplot2: Create Elegant Data Visualisations Using the Grammar of Graphics*. R package version 3.1.1.

Wickham, H. and Grolemund, G. (2017). *R for Data Science*. O'Reilly, New York.

Wilkinson, L. (2012). The grammar of graphics. In *Handbook of Computational Statistics*, pages 375–414. Springer.

Xavier, S.-i.-M. et al. (1997). I just ran two million regressions. *American Economic Review*, 87(2):178–83.

Xie, Y. (2019). *bookdown: Authoring Books and Technical Documents with R Markdown*. R package version 0.10.

Index